Leander Miller Hoskins

The Elements of Graphic Statics

A Text-Book for Students of Engineering

Leander Miller Hoskins

The Elements of Graphic Statics
A Text-Book for Students of Engineering

ISBN/EAN: 9783337214609

Printed in Europe, USA, Canada, Australia, Japan

Cover: Foto ©berggeist007 / pixelio.de

More available books at **www.hansebooks.com**

ELEMENTS OF GRAPHIC STATICS.

THE

ELEMENTS OF GRAPHIC STATICS

A Text-Book for Students of Engineering

BY

L. M. HOSKINS

PROFESSOR OF APPLIED MECHANICS IN THE LELAND STANFORD
JUNIOR UNIVERSITY

REVISED EDITION

New York
THE MACMILLAN COMPANY
LONDON: MACMILLAN & CO., LTD.
1899

Norwood Press
J. S. Cushing & Co. — Berwick & Smith
Norwood Mass. U.S.A.

PREFACE TO REVISED EDITION.

THE method of treatment adopted in this work is designed to meet the needs of the beginner. To this end the endeavor has been made to secure simplicity of presentation without sacrifice of logical rigor.

In scope, the work has been planned with reference to the requirements of students of engineering. This limits the development of the general theory to such principles and methods as are practically useful. It also excludes many applications which, though leading to practical results, are likely to prove useful and to save labor only in the hands of the expert in graphical methods. Graphic Statics is treated as a branch of Mechanics rather than of Geometry, and those beautiful developments whose chief interest is geometrical have not been included.

Although graphical methods are especially useful to the structural engineer, it is believed that students in all departments of engineering will find it profitable to become familiar with the general theory of complanar forces from the graphical side, as well as with the simpler applications to the determination of stresses in framed structures, and of centroids and moments of inertia of plane areas. The application to trusses carrying moving loads is of less general interest, its practical utility being limited mainly to the solution of problems in bridge design. The methods developed in Chapter VII will therefore be of practical value mainly to students giving special attention to this branch of engineering.

In the present revised edition no change has been made in general plan, and few changes in the treatment adopted, except in the portions relating to beams and trusses carrying moving loads (Chapters VI and VII). These portions have been wholly re-written. It is believed that a substantial improvement has been made upon the methods hitherto used, particularly in the criterion for determining the position of a given load-series which causes maximum stress in any member of a truss. The improvement consists in generalization, which is believed to be gained without sacrifice of simplicity. The graphical method of applying the criterion in the case of trusses with parallel chords has been fully treated by Professor H. T. Eddy. The method here given applies without the restriction to parallel chords. The algebraic statement of the same criterion, as given in Art. 152, is also believed to be a useful generalization of the methods heretofore used. Whether the algebraic or the graphical treatment is preferred, a method is useful in proportion to its generality, provided this does not involve a loss of simplicity. There is a decided advantage in the use of a single general equation, applicable to any member of any truss, instead of several particular equations, each applicable to a special member or to a special form of truss.

STANFORD UNIVERSITY, CALIFORNIA,
November, 1898.

CONTENTS.

PART I.—GENERAL THEORY.

CHAPTER I. DEFINITIONS.—CONCURRENT FORCES.

		PAGE
§ 1.	Preliminary Definitions	1
2.	Composition of Concurrent Forces	5
3.	Equilibrium of Concurrent Forces	6
4.	Resolution of Concurrent Forces	8

CHAPTER II NON-CONCURRENT FORCES.

§ 1.	Composition of Non-concurrent Forces Acting on the Same Rigid Body	11
2.	Equilibrium of Non-concurrent Forces	18
3.	Resolution into Non-concurrent Systems	28
4.	Moments of Forces and of Couples	30
5.	Graphic Determination of Moments	35
6.	Summary of Conditions of Equilibrium	37

CHAPTER III. INTERNAL FORCES AND STRESSES.

§ 1.	External and Internal Forces	40
2.	External and Internal Stresses	42
3.	Determination of Internal Stresses	46

PART II.—STRESSES IN SIMPLE STRUCTURES.

CHAPTER IV. INTRODUCTORY.

§ 1.	Outline of Principles and Methods	53

CHAPTER V. ROOF TRUSSES. — FRAMED STRUCTURES SUSTAINING STATIONARY LOADS.

§ 1. Loads on Roof Trusses 58
2. Roof Truss with Vertical Loads 61
3. Stresses Due to Wind Pressure 67
4. Maximum Stresses 71
5. Cases Apparently Indeterminate 74
6. Three-hinged Arch 80
7. Counterbracing 88

CHAPTER VI. SIMPLE BEAMS.

§ 1. General Principles 94
2. Beam Sustaining Fixed Loads 98
3. Beam Sustaining Moving Loads 101

CHAPTER VII. TRUSSES SUSTAINING MOVING LOADS.

§ 1. Bridge Loads 112
2. Truss Regarded as a Beam 115
3. Truss Sustaining Any Series of Moving Loads 120
4. Truss with Subordinate Bracing 137
5. Uniformly Distributed Moving Load 145

PART III. — CENTROIDS AND MOMENTS OF INERTIA.

CHAPTER VIII. CENTROIDS.

§ 1. Centroid of Parallel Forces 152
2. Center of Gravity — Definitions and General Principles . . . 157
3. Centroids of Lines and of Areas 160

CHAPTER IX. MOMENTS OF INERTIA.

§ 1. Moments of Inertia of Forces 167
2. Moments of Inertia of Plane Areas 177

CHAPTER X. CURVES OF INERTIA.

§ 1. General Principles 187
2. Inertia-Ellipses for Systems of Forces 190
3. Inertia-Curves for Plane Areas 195

GRAPHIC STATICS.

PART I.

GENERAL THEORY.

CHAPTER I. — DEFINITIONS. CONCURRENT FORCES.

§ 1. *Preliminary Definitions.*

1. **Dynamics** treats of the action of forces upon bodies. Its two main branches are Statics and Kinetics.

Statics treats of the action of forces under such conditions that no change of motion is produced in the bodies acted upon.

Kinetics treats of the laws governing the production of motion by forces.

2. **Graphic Statics** has for its object the deduction of the principles of statics, and the solution of its problems, by means of geometrical figures.

3. **A Force** is that which tends to change the state of motion of a body. We conceive of a force as a *push* or a *pull* applied to a body at a definite point and in a definite direction. Such a push or pull *tends* to give motion to the body, but this tendency may be neutralized by the action of other forces. The effect of a force is completely determined when three things are given, — its *magnitude*, its *direction*, and its *point of application*. The line parallel to the direction of the force and containing its point of application, is called its *line of action*.

Every force acting upon a body is exerted *by some other body*. But the problems of statics usually concern only the body *acted upon*. Hence, frequently, no reference is made to the bodies exerting the forces.

4. Unit Force. — The *unit force* is a force of arbitrarily chosen magnitude, in terms of which forces are expressed. Several different units are in use. The one employed in this work is the *pound*, which will now be defined.

A *pound force* is a force equal to the weight of a pound mass at the earth's surface. A *pound mass* is the quantity of matter contained in a certain piece of platinum, arbitrarily chosen, and established as the standard by act of the British Parliament.

The pound force, as thus defined, is not perfectly definite, since the weight of any given mass (that is, the attraction of the earth upon it) is not the same for all positions on the earth's surface. The variations are, however, unimportant for most of the requirements of the engineer.

In its fundamental meaning, the word "pound" refers to the unit mass, and it is unfortunate that it is also applied to the unit force. The usage is, however, so firmly established that it will be here followed.

5. Concurrent and Non-concurrent Forces. — Forces acting on the same body are *concurrent* when they have the same point of application. When applied at different points they are non-concurrent.

6. Complanar Forces are those whose lines of action are in the same plane. In this work, only complanar systems will be considered unless otherwise specified.

7. A Couple is the name given to a system consisting of two forces, equal in magnitude, but opposite in direction, and having different lines of action. The perpendicular distance between the two lines of action is called the *arm* of the couple.

8. **Equivalent Systems of Forces.** — Two systems of forces are *equivalent* when either may be substituted for the other without change of effect.

9. **Resultant.** — A single force that is equivalent to a given system of forces is called the *resultant* of that system. It will be shown subsequently that a system of forces may not be equivalent to any single force. When such is the case, the simplest system equivalent to the given system may be called its resultant. Any forces having a given force for their resultant are called *components* of that force.

10. **Composition and Resolution of Forces.** — Having given any system of forces, the process of finding an equivalent system is called the *composition of forces*, if the system determined contains fewer forces than the given system. If the reverse is the case, the process is called the *resolution of forces*.

The process of finding the resultant of any given forces is the most important case of *composition;* while the process of finding two or more forces, which together are equivalent to a single given force, is the most common case of *resolution.*

11. **Representation of Forces Graphically.** — The *magnitude* and *direction* of a force can both be represented by a line; the length of the line representing the magnitude of the force, and its direction the direction of the force.

In order that the length of a line may represent the magnitude of a force, a certain length must be chosen to denote the unit force. Then a force of any magnitude will be represented by a length which contains the assumed length as many times as the magnitude of the given force contains that of the unit force.

In order that the direction of a force may be represented by a line, there must be some means of distinguishing between the two opposite directions along the line. The usual method is to place an arrow-head on the line, pointing in the direction toward which the force acts. If the line is designated by letters placed at its extremities, the order in which these are read may indicate the direction of the force. Thus, *AB* and *BA* represent two forces, equal in magnitude but opposite in direction.

The *line of action* of a force can also be represented by a line drawn on the paper.

In solving problems in statics, it is usually convenient to draw two separate figures, in one of which the forces are represented in *magnitude and direction* only, and in the other in *line of action* only.

These two species of diagrams will be called *force diagrams* and *space diagrams*, respectively.

12. Notation. — The use of graphic methods is much facilitated by the adoption of a convenient system of notation in the figures drawn.

There will generally be two figures (the force diagram and the space diagram) so related that for every line in one there is a corresponding line in the other.

In the force diagram each line represents a force in *magnitude* and *direction;* in the space diagram the corresponding line represents the *line of action* of the force. These lines will usually be designated in the following manner: The line denoting the magnitude and direction of the force will be marked by two capital letters, one at each extremity; while the action-line will be marked by the corresponding small letters, one being placed at each side of the line designated. Thus, in Fig. 1, *AB* represents a force in magnitude and direction, while its action-line is marked by the letters *ab*, placed as shown.

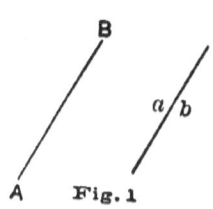
Fig. 1

§ 2. Composition of Concurrent Forces.

13. **Resultant of Two Concurrent Forces.** — If two concurrent forces are represented in magnitude and direction by two lines AB and BC, their resultant is represented in magnitude and direction by AC. (Fig. 2.) Proofs of this proposition are given in all elementary treatises on mechanics, and the demonstration will be here omitted. The *point of application* of the resultant is the same as that of the given forces. Thus if O (Fig. 2) is the given point of application, then ab, bc, and ac, drawn parallel to AB, BC, and AC respectively, are the lines of action of the two given forces and their resultant. The figure marked (A) is a *force diagram*, and (B) is the corresponding *space diagram* (Art. 11).

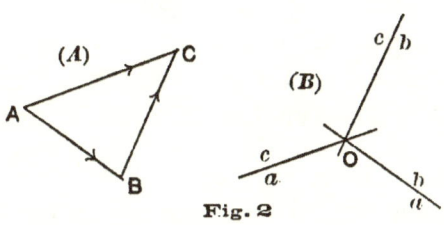

Fig. 2

14. **Resultant of Any Number of Concurrent Forces.** — If any number of concurrent forces are represented in magnitude and direction by lines AB, BC, CD, . . ., their resultant is represented in magnitude and direction by the line AN, where N is the extremity of the line representing the last of the given forces.

This proposition follows immediately from the preceding one; for the resultant of the forces represented by AB and BC is a force represented by AC; the resultant of AC and CD is AD, and so on. By continuing the process we shall arrive at the result stated. It is

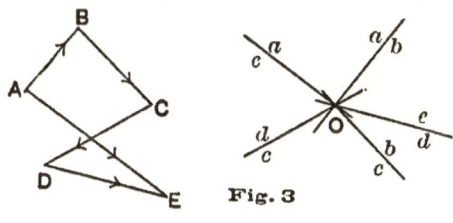

Fig. 3

readily seen that the order in which the forces are taken does not affect the magnitude or direction of the resultant as thus determined. The point of application of the resultant is the same as that of the given forces. Figure 3 shows the force diagram and space diagram for a system of four forces represented by AB, BC, CD, DE, and their resultant represented by AE, applied at the point O.

From the above construction it is evident that every system of concurrent forces has for its resultant some single force (Art. 9); though in particular cases its magnitude may be zero.

15. **Force Polygon.** — The figure formed by drawing in succession lines representing in magnitude and direction any number of forces is called a *force polygon* for those forces. Thus Fig. 4 is a force polygon for any four forces represented in magnitude and direction by the lines AB, BC, CD, and DE, whatever their lines of action. It may happen that the point E coincides with A, in which case the polygon is said to be *closed*. It is evident that the order in which the forces are taken does not affect the relative positions of the initial and final points.

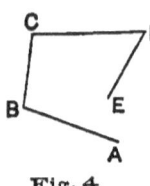
Fig. 4.

§ 3. *Equilibrium of Concurrent Forces.*

16. **Definition.** — A system of forces acting on a body is in *equilibrium* if the motion of the body is unchanged by its action.

17. **Condition of Equilibrium.** — In order that no motion may result from the action of any system of concurrent forces, the magnitude of the resultant must be zero; and conversely, if the magnitude of the resultant is zero, no motion can result. But (Arts. 14 and 15) the condition that the resultant is zero is identical with the condition that the force polygon closes. Hence, the following proposition:

If any system of concurrent forces is in equilibrium, the force polygon for the system must close. And conversely, *If the force polygon is closed for any system of concurrent forces, the system is in equilibrium.*

The comparison of this with the analytical conditions of equilibrium is given in Art. 22.

18. **Method of Solving Problems in Equilibrium.** — If a system of concurrent forces in equilibrium be partially unknown, we may in certain cases determine the unknown elements by applying the principles of Art. 17.

A common case is that in which two forces are unknown except as to lines of action. Thus, suppose a system of five forces in equilibrium, three being fully known, represented in magnitude and direction by AB, BC, CD (Fig. 5), and in lines of action by ab, bc, cd,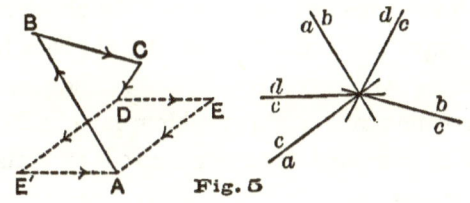
while concerning the other two we know only their lines of action dc, ea. To determine these two in magnitude and direction, it is necessary only to complete the force polygon of which $ABCD$ is the known part. The remaining sides must be parallel respectively to dc and ea. From D draw a line parallel to dc, and from A a line parallel to ea, prolonging them till they intersect at E. Then DE and EA represent the required forces in magnitude and direction, and the complete force polygon is $ABCDEA$. It is evident that $ABCDE'A$ is an equally legitimate form of the force polygon, and gives the same result for the magnitude and direction of each of the unknown forces. This problem occurs constantly in the construction of stress diagrams by the method described in Part II.

The student will find little difficulty in treating other problems in the equilibrium of concurrent forces.

19. **Problems in Equilibrium.** — (1) A particle is in equilibrium under the action of five forces, three of which are

completely known, while of the remaining two, one is known in direction only, and the other in magnitude only. To determine the unknown forces.

(2) Suppose two forces known in magnitude but not in direction, the remaining forces being wholly known.

(3) Suppose one force wholly unknown, the others being known.

§ 4. *Resolution of Concurrent Forces.*

20. **To Resolve a Given Force into Any Number of Components** having the same point of application, we have only to draw a closed polygon of which one side shall represent the magnitude and direction of the given force; then the remaining sides will represent, in magnitude and direction, the required components. This problem is, in general, indeterminate, unless the components are required to satisfy certain specified conditions.

[NOTE. — A problem is said to be *indeterminate* if its conditions can be satisfied in an infinite number of ways. It is *determinate* if it admits of only one solution. Thus, the problem, to determine the values of x and y which shall satisfy the equation $x + y = 10$, is indeterminate; while the problem, given $2x + 3 = 7$, to find the value of x, is determinate. The case in which a finite number of solutions is possible may be called *incompletely determinate*. Thus, the problem, given $x^2 + x - 6 = 0$, to find x, admits of two solutions, and therefore is incompletely determinate. All these classes of problems may be met with in statics.]

21. **To Resolve a Given Force into Two Components.** — This problem is indeterminate unless additional data are given. For if the given force be represented in magnitude and direction by a line, any two lines which with the given line form a triangle may represent forces which are together equivalent to the given force. But an infinite number of such triangles may be drawn. The solution of the following four cases of this problem will form exercises for the student. In each case the *force diagram* and *space diagram* should be completely drawn, and the student should notice whether the problem is determinate, partially determinate, or indeterminate.

RESOLUTION OF CONCURRENT FORCES.

(1) Let the lines of action of the required components be given.

(2) Let the two components be given in magnitude only.

(3) Let the line of action of one component and the magnitude of the other be given.

(4) Let the magnitude and direction of one component be given.

It will be noticed that these four cases correspond to four cases of the solution of a plane triangle.

22. Resolved Part of a Force. — If a force is conceived to be replaced by two components at right angles to each other, each is called the *resolved part*,* in its direction, of the given force.

It is readily seen that the resolved part of a force represented by AB (Fig. 6) in the direction of any line XX, is represented in magnitude and direction by $A'B'$, the orthographic projection of AB upon XX. It follows that the resolved part (in any given direction) of the resultant of any concurrent forces is equal to the algebraic sum of the resolved parts of its components in that direction; signs plus and minus being given to the resolved parts to distinguish the two opposite directions which they may have. Thus (Fig. 7) the resolved parts of the forces AB, BC, CD, in a direction parallel to XX, are $A'B'$, $B'C'$, $C'D'$; and their algebraic sum is $A'D'$, which is the resolved part of the resultant AD.

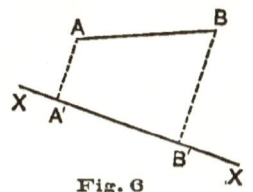

Fig. 6

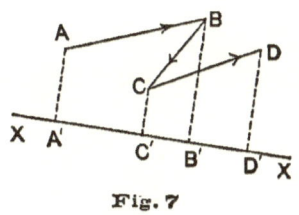

Fig. 7

If the resultant is zero, D' coincides with A'; hence,

(1) For the equilibrium of any concurrent forces, the sum of their resolved parts in any direction must be zero.

---

* The term "resolute" has been proposed by J. B. Lock ("Elementary Statics") to denote what is here defined as the *resolved part* of a force.

Again, if D' coincides with A', then either D coincides with A, or else AD is perpendicular to XX; hence,

(2) If the sum of the resolved parts of any concurrent forces in a given direction is zero, their resultant (if any) is perpendicular to that direction. And if the sum of the resolved parts is zero for each of two directions, the resultant is zero, and the system is in equilibrium.

Propositions (1) and (2) state the conditions of equilibrium for concurrent forces usually deduced in treatises employing algebraic methods.

CHAPTER II.—NON-CONCURRENT FORCES.

§ 1. *Composition of Non-concurrent Forces Acting on the Same Rigid Body.*

23. **Definition of Rigid Body.**—A *rigid body* is one whose particles do not change their positions relative to each other under any applied forces. No known body is perfectly rigid, but for the purposes of statics, most solid bodies may be considered as such; and any body which has assumed a form of equilibrium under applied forces, may, for the purposes of statics, be treated as a rigid body without error.

24. **Change of Point of Application.**—The effect of a force upon a rigid body will be the same, at whatever point in its line of action it is applied, if only the particle upon which it acts is rigidly connected with the body.

This proposition is fundamental to the development of the principles of statics, and is amply justified by experience.*
In applying the principle, we are at liberty to assume a point of application outside the actual body, the latter being ideally extended to any desired limits.

25. **Resultant of Two Non-Parallel Forces.**—If two complanar forces are not parallel, their lines of action must intersect, and the point of intersection may be taken as the point of application of each force. Hence, they may be treated as

---

* This proposition may be proved analytically by deducing the equations of motion of a rigid body, and showing that the effect of any force on the motion of the body depends only upon its magnitude, direction, and line of action. But such a proof is, of course, outside the scope of this work.

concurrent forces, and their resultant may be determined as in Art. 13. The following proposition may therefore be stated:

If two forces acting in the same plane on a rigid body are represented in magnitude and direction by AB and BC, their resultant is represented in magnitude and direction by AC, and its line of action passes through the point of intersection of the lines of action of the given forces. Its point of application may be any point of this line.

It may happen that the point of intersection of the two given lines of action falls outside the limits available for the drawing. In such a case it will be most convenient to find the resultant by the method to be explained in Art. 27. The same remark applies to the case of two parallel forces.

26. Resultant of Any Number of Non-concurrent Forces — First Method. — The method of the preceding article may be extended to the determination of the resultant of any number of forces acting on the same rigid body. Let AB, BC, CD, DE (Fig. 8), represent in magnitude and direction four forces, and let ab, bc, cd, de represent their lines of action. To find their resultant, we may proceed as follows: The resultant of AB and BC is represented

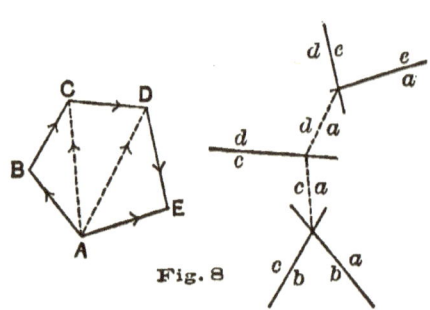

Fig. 8

in magnitude and direction by AC, and in line of action by ac drawn parallel to AC through the point of intersection of ab and bc. Combining this resultant with CD, we get as *their* resultant a force represented in magnitude and direction by AD, and in line of action by ad drawn parallel to AD through the point of intersection of ac and cd. This is evidently the resultant of AB, BC, and CD. In the same way, this resultant

COMPOSITION OF NON-CONCURRENT FORCES. 13

combined with DE gives for *their* resultant a force whose magnitude and direction are represented by AE, and whose line of action is ac, parallel to AE and passing through the point in which ad intersects de. This last force is the resultant of the four given forces.

The process may evidently be extended to the case of any number of forces.

As in the case discussed in the preceding article, this method will become inapplicable or inconvenient if any of the points of intersection fall outside the limits available for the drawing. For this reason it is usually most convenient to employ the method described in Art. 27.

The student should bear in mind that the length and direction AE and the line ac are not the magnitude, direction, and line of action of any actual force applied to the body. By the resultant is meant an ideal force, which, if it acted, would produce the same effect upon the motion of the body as is produced by the given forces. It is a force which may be conceived to replace the actual forces, and may be assumed to be applied to any particle in its line of action, provided that particle is regarded as rigidly connected with the given body. The line of action may in reality fail to meet the given body. (See Art. 24.)

27. **Resultant of Non-concurrent Forces — Second Method.** — This method will be described by reference to an example. Referring to Fig. 9, let AB, BC, CD, DE represent in magnitude and direction four forces whose lines of action are ab, bc, cd, de; and let it be required to find their resultant. Draw the force polygon $ABCDE$, and from any point O in the force diagram draw lines OA, OB, OC, OD, OE. These lines may represent, in magnitude and direction, components into which the given forces may be resolved. Thus AB is equivalent to forces represented by AO and OB acting in any lines parallel to AO, OB, whose point of intersection falls upon ab; BC is

equivalent to forces represented by BO, OC, acting in any lines parallel to BO, OC, which intersect on bc; and so for each of the given forces. The four given forces may, therefore, be replaced by eight forces given in magnitude and direction by

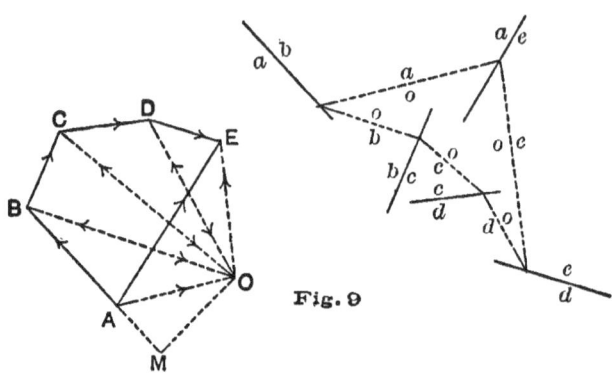

Fig. 9

AO, OB, BO, OC, CO, OD, DO, OE, with proper lines of action. Now, it is possible to make the lines of action of the forces represented by OB and BO coincide; and the same is true of each pair of equal and opposite forces, OC, CO; OD, DO. To accomplish this, let AO, OB act in lines ao, ob, intersecting at any assumed point of ab. Prolong ob to intersect bc, and take the point thus determined as the point of intersection of the lines of action of BO, OC; these lines are then bo, oc. Similarly prolong oc to intersect cd, and let the point of intersection be taken as the point at which CD is resolved into CO and OD; the lines of action of these forces are then co and od. In like manner choose do, oe, intersecting on de, as the lines of action of DO, OE. If this is done, the forces OB, BO will neutralize each other and may be omitted from the system; also the pairs OC, CO, and OD, DO. Hence, there remain only the two forces represented in magnitude and direction by AO, OE, and in lines of action by ao, oe. Their resultant is given in magnitude and direction by AE, and its line of action is ae, drawn parallel to AE through the

COMPOSITION OF NON-CONCURRENT FORCES. 15

point of intersection of *oa* and *oe*; and this is also the resultant of the given system.

By carefully following through this construction the student will be able to reduce it to a mechanical method, which can be readily applied to any system.

28. Funicular Polygon. — The polygon whose sides are *oa*, *ob*, *oc*, *od*, *oe*, is called a *funicular polygon* * for the given forces.

Since the point at which the two components of *AB* are assumed to act may be taken anywhere on the line *ab*, there may be any number of funicular polygons with sides parallel to *oa*, *ob*, etc. Again, if *O* is taken at a different point, there may be drawn a new funicular polygon starting at any point of *ab*; and by changing the starting point any number of funicular polygons may be drawn with sides parallel to the new directions of *OA*, *OB*, etc. Moreover, different force and funicular polygons may be obtained by changing the order in which the forces are taken.

It may be proved geometrically that for every possible funicular polygon drawn for the same system of forces, the last vertex, determined by the above method, will lie on the same line parallel to the closing side of the force polygon (as *ae*, Fig. 9). Such a proof is outside the scope of this work. The truth of the proposition may, however, be shown from the principles of statics. For if it were not true, it would be possible by the above method to find two or more forces, *having different lines of action*, which are equivalent to each other, because each is equivalent to the given system. But this is impossible.

29. Examples. — 1. Choose five forces, assigning the magnitude, direction, and line of action of each, and find their resultant by constructing the force and funicular polygons.

2. Draw a second funicular polygon, using the same point *O* in the force diagram.

---

*Also called equilibrium polygon.

3. Draw a third funicular polygon, choosing a new point O.

4. Solve the same problem, taking the forces in a different order.

30. **Definitions.** — The point O (Fig. 9) is called the *pole* of the force polygon. The lines drawn from the pole to the vertices of the force polygon may be called *rays*. The sides of the funicular polygon are sometimes called *strings*.

Each *ray* in the force diagram is parallel to a corresponding *string* in the space diagram. As a mechanical rule, it should be remembered that *the two rays drawn to the extremities of the line representing any force are respectively parallel to the two strings which intersect on the line of action of that force.*

The rays terminating at the extremities of any side of the force polygon represent in magnitude and direction two components that may replace the force represented by that side; while the corresponding strings represent the lines of action of these components. Thus (Fig. 9) the force represented by BC may be replaced by two forces represented by BO, OC, acting in the lines bo, oc.

The *pole distance* of any force is the perpendicular distance from the pole to the line representing that force in the force diagram. It may evidently be considered as representing the resolved part, perpendicular to the given force, of either of the components represented by the corresponding rays. Thus, OM (Fig. 9) is the pole distance of AB; and OM represents the resolved part, perpendicular to AB, of either OA or OB. The student should notice particularly that the pole distance represents a *force magnitude* and not a length.

31. **Forces not Possessing a Single Resultant.** — It may happen that the first and last sides of the funicular polygon are parallel, so that the above construction fails to give the line of action of the resultant. This will be the case if the pole is chosen on the line AE (Fig. 9), because the first and last sides of the funicular polygon are respectively parallel to OA and

COMPOSITION OF NON-CONCURRENT FORCES. 17

OE. The difficulty will, in this case, be avoided by taking the pole at some point not on the line *AE*. But in one particular case *OA* and *OE* will be parallel, *wherever the pole be taken*. By inspection of the force diagram it is seen that this can occur only when the point *E* coincides with *A*. In this case *AO* and *OE* represent equal and opposite forces; and unless their lines of action coincide, they cannot be combined into a simpler system. If their lines of action are coincident, the forces neutralize each other and the resultant is zero; if not, the system reduces to a *couple* (Art. 7). If the lines of action of *AO* and *OE* coincide, they may still be regarded as forming a couple, its arm being zero. Therefore,

If the force polygon for any system of forces closes, the resultant is a couple.

Now, it is evident that by shifting the starting point for the funicular polygon, the lines of action of *AO* and *OE* will be shifted; and by taking a new pole, their magnitude, or direction, or both, may be changed. Hence, there may be found any number of couples, each equivalent to the same system of forces, and therefore equivalent to each other. The relation between equivalent couples is discussed in Art. 52.

Example. — Assume five forces whose force polygon closes, the lines of action being taken at random. Draw two funicular polygons, using the same pole, and a third, using a different pole; thus reducing the given system to an equivalent couple in three different ways.

32. **Resultant Force and Resultant Couple.** — From the above discussion (Arts. 27 and 31), it is evident that any system of complanar forces, acting on the same rigid body, is equivalent either to a single force or to a couple. In other words, *every system of complanar forces possesses either a resultant force or a resultant couple*. (See Art. 9.)

33. **Comparison of Methods.** — Of the methods given in Arts. 26 and 27, for finding the resultant of a system of non-concur-

rent forces, the first is a special case of the second. For if the pole in Fig. 9 be chosen at the point A, and the first string be made to pass through the point of intersection of ab and bc, the construction becomes identical with that of Fig. 8. If the first method is employed, we are liable to meet the difficulty mentioned in Arts. 25 and 26, that some of the required points of intersection do not fall within convenient limits.

In the second method, since the pole may be chosen at pleasure, the rays may generally be caused to make convenient angles with the given forces, and all required points of intersection to fall within convenient limits.

34. **Closing of the Funicular Polygon.** — Let the force polygon be drawn for any given forces, and let A and E be the initial and final points. Suppose a funicular polygon drawn, corresponding to any pole O. The given system is equivalent to two forces represented in magnitude and direction by AO, OE, their lines of action being the first and last sides of the funicular polygon. In general, these lines of action will not be parallel; but, as explained in Art. 31, it may happen that they are parallel. And if parallel, it may happen that they coincide. In this case, the funicular polygon is said to be *closed*.

§ 2. *Equilibrium of Non-concurrent Forces.*

35. **Conditions of Equilibrium.** — From the foregoing principles the conditions of equilibrium for any system of complanar forces may be deduced.

Whatever the system of forces may be, let a force polygon be drawn, and let A and E be its initial and final points. Choose a pole O, and draw a funicular polygon. The system is thus reduced to two forces represented in magnitude and direction by AO and OE; their lines of action being the first and last sides of the funicular polygon. In order that there may be equilibrium, these two forces must be equal and opposite and have the same line of action. In order that they may be equal

EQUILIBRIUM OF NON-CONCURRENT FORCES.

and opposite, A and E must coincide. In order that they may have the same line of action, the first and last sides of the funicular polygon must coincide. That is, in order that there may be equilibrium, two conditions must be satisfied:

(*a*) *The force polygon must close.*
(*b*) *The funicular polygon must close.*

Conversely, *if the force polygon closes and one funicular polygon also closes, the system is in equilibrium.* For, if the force polygon closes, the two forces AO and OE are equal and opposite; and if a funicular polygon closes, these two forces have the same line of action and therefore balance each other.

It follows that if the force polygon and one funicular polygon are closed, all funicular polygons will close.

36. **Auxiliary Conditions of Equilibrium.** — If any system of forces in equilibrium be divided into two groups, the resultants of the two groups are equal and opposite and have the same line of action.

Particular case. — If three forces are in equilibrium, their lines of action must meet in a point, or be parallel. For, if the lines of action of two of the forces intersect, their resultant must act in a line passing through the point of intersection. But this resultant must be equal and opposite to the third force and have the same line of action.

37. **General Method of Solving Problems in Equilibrium.** — The principles of Art. 35 furnish a general method of solving, graphically, problems relating to the equilibrium of complanar forces acting on a rigid body. The problems to be solved will always be of the following kind:

A body is in equilibrium under the action of forces, some of which are completely known, and others wholly or partly unknown. It is required to determine the unknown elements. The required elements may be either the magnitudes or the lines of action of the forces.

GRAPHIC STATICS.

The general method of procedure is as follows: First, draw the force polygon and funicular polygon so far as possible from the given data; then complete them, subject to the condition that both polygons must close.

This general method will be illustrated by the solution of several problems of frequent occurrence. We may here meet with both determinate and indeterminate problems (Art. 20). In order that a problem may be determinate, the given data, together with the condition that the force polygon and funicular polygon are to close, must be just sufficient to fully determine these figures. There are many possible cases furnishing determinate problems. Some of these cannot readily be solved graphically. In the following articles are treated three important cases to which the general method above outlined is well adapted.

38. **Problems in Equilibrium.** I. — A rigid body is in equilibrium under the action of a system of parallel forces, all known except two, these being unknown in magnitude and direction, but having known lines of action. It is required to fully determine the unknown forces.

Let the known lines of action of five forces in equilibrium be ab, bc, cd, de, ca (Fig. 10), and let AB, BC, CD represent the known forces in magnitude and direction, while DE and EA are at first unknown. The force polygon, so far as it can be drawn from the given data, is the straight line $ABCD$. Choose a pole O and draw rays OA, OB, OC, OD. Parallel to these draw in the space diagram the strings oa, ob, oc, od, four successive sides of the funicular polygon. This much can be drawn from the data given. We must now complete the

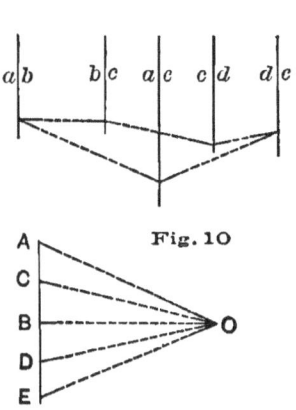

Fig. 10

two polygons and cause both to close. In the funicular polygon, but one side remains to be drawn; and in the force polygon but one vertex remains unknown. If the forces are taken in the order AB, BC, CD, DE, EA, the unknown vertex in the force polygon is the one to be marked E; the unknown side of the funicular polygon is oe, and is to be parallel to the line OE. But the string oe must pass through the intersection of od with de, and also through the intersection of oa with ae. Hence, since these intersections are both known, oe can be drawn at once as shown in the figure; and then OE can be drawn parallel to oe. This fixes the point E; and DE and EA represent, in magnitude and direction, the required forces whose lines of action are de, ea.

39. **Problems in Equilibrium. II.** — A rigid body is in equilibrium under the action of a system of non-parallel forces, all known, except two; of these, the line of action of one and the point of application of the other are given. It is required to completely determine the unknown forces.

Let the system consist of five forces to be represented in the usual way; in magnitude and direction by lines marked AB, BC, CD, DE, EA; and in lines of action by ab, bc, cd, de, ca. Of these lines let bc, cd, de, ea (Fig. 11), be given, and let L be a given point of ab. Also let BC, CD, and DE be known, while EA, AB are unknown. First, draw the force polygon as far as possible, giving B, C, D, E, four consecutive vertices of the force polygon. The side EA must be parallel to ca, but its length is unknown, hence the vertex A of the force polygon cannot be fixed. Next, draw the funicular polygon so

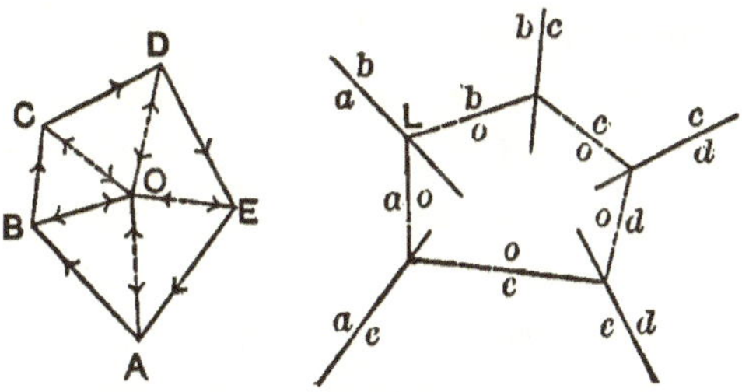

Fig. 11

far as possible from the given data. Choose the pole O, and draw the rays OB, OC, OD, OE. The remaining ray, OA, cannot yet be drawn. Now, in the funicular polygon, the sides ob, oc must intersect on bc; oc and od must intersect on cd; od and oe must intersect on de; oe and oa must intersect on ea; and oa and ob must intersect on ab. Since L is the only known point of the line ab, let this point be taken as the point of intersection of oa and ob. Now draw ob through L parallel to OB, and successively oc, od, oe, parallel to OC, OD, OE. We may now draw oa, joining L with the point in which oe intersects ea. This completes the funicular polygon. The force polygon can now be completed as follows: From O draw a line parallel to oa and from E a line in the known direction of EA; their intersection determines A. This determines both EA and AB, and the force polygon is completely known. The line of action ab may now be drawn through L parallel to AB, and the forces are completely determined.

40. **Problems in Equilibrium.** III. — A rigid body is in equilibrium under the action of any number of forces, of which three are known only in line of action; the remaining forces being completely known. It is required to determine the unknown forces.

Let the given forces be six in number, their lines of action being represented by ab, bc, cd, de, ef, fa (Fig. 12), and let AB, BC, and CD be known, while the remaining three forces are unknown in magnitude and direction. The known data make it possible to draw at once three sides of the force polygon, namely, AB, BC, CD; and the four rays OA, OB, OC, OD, from any assumed pole O. Also, four sides of the funicular polygon, oa, ob, oc, od, may be drawn at once. But oe and of are unknown; also DE, EF, and FA in the force polygon.

Since any side of the funicular polygon can be drawn through any chosen point, let the polygon be started by drawing oa

through the intersection of *ef* and *fa*. We may then draw successively *ob, oc, od*. Now, *of* is unknown in direction; but it is to be drawn through the intersection of *oa* and *fa*, hence, whatever its direction, it intersects *ef* in the same point (since *oa* was drawn through the intersection of *ef* and *fa*). Hence, the two vertices of the funicular polygon falling on *ef* and *fa* coincide at the intersection of these two lines. We may there-

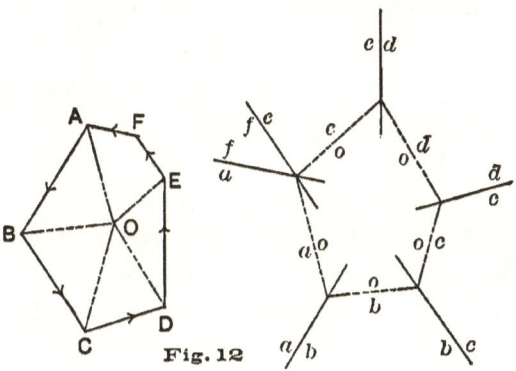

Fig. 12

fore draw *oe* through this point and also through the point already found by the intersection of *od* and *de*. Now draw a line from *O* parallel to *oc*, and from *D* a line parallel to *de*; their intersection is *E*. Draw from *E* a line parallel to *ef*, and from *A* a line parallel to *fa*; their intersection determines *F*. The force polygon is now completely drawn, and *DE, EF, FA* represent, in magnitude and direction, the required forces. The remaining ray *OF* and the corresponding string *of* may now be drawn, but are not needed.

The construction might have been made equally well by choosing the intersection of *de* and *ef* as the starting point, since two vertices of the funicular polygon may be made to coincide at that point. Or, the point of intersection of *de* and *af* might be chosen; but in that case, the order of the forces should be so changed as to make *de* and *af* consecutive. If this were done the figures should be relettered to agree with

GRAPHIC STATICS.

the order in which the forces were taken. It may be noticed that the *direction* of the string first drawn may be chosen arbitrarily and the pole so taken as to correspond to the direction chosen. This is important in the treatment of the following special case.

Case of inaccessible points of intersection. — It may happen that the lines of action *de*, *ef*, and *fa* have no point of intersection within convenient limits. When this is the case, the method just given may still be applied, but involves the geometrical problem of drawing a line through an inaccessible point. For example, if *ef* and *fa* intersect beyond the limits of the drawing, as shown in Fig. 13, we may proceed as follows: Choose some point of *ab* and from it draw a line through the point of intersection of *ef* and *fa*. (This can be done by a method to be explained presently.) Let this line be *oa*. From the known point *A* of the force polygon draw a line parallel to

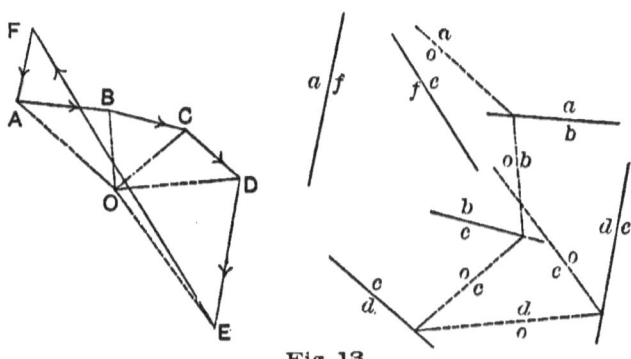

Fig. 13

oa and choose a pole upon it. Draw the rays *OB*, *OC*, *OD*; then the corresponding strings in the order *ob*, *oc*, *od*. From the point in which *od* intersects *de* draw a line to the inaccessible point of intersection of *ef* and *fa*; this will be *oe*. The force polygon can now be completed just as in the preceding case.

A line may be drawn through the inaccessible point of intersection of two given lines by the following method: Let *AC*,

BD (Fig. 14), be the given lines, and let it be required to draw a line through their point of intersection from some point *P*. Draw *PA* intersecting *AC* and *BD*, and from *C*, any point of *AC*, draw a line *CQ* parallel to *PA*. From *E*, a point of *BD*, draw *EA*, *EP*; also draw *CF* parallel to *AE*, and *FQ* parallel to *EP*. Then *PQ* is the line required. For, by the similar triangles, it is easily shown that

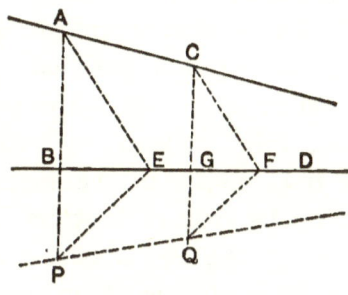

Fig. 14

$\dfrac{AB}{BP} = \dfrac{CG}{GQ}$; which proves that *AC*, *BG*, and *PQ* meet in a point.

Exception. — If the lines of action of the three unknown forces meet in a point (or are parallel), the problem is impossible of solution, unless it happens that the resultant of the known forces acts in a line through the point of intersection of those three lines of action (or parallel to them); in which case the problem is indeterminate. These cases will not be discussed here.

Remark. — In the problems treated in this and the two preceding articles, it will be noticed that the forces should be taken in such an order that those which are completely known are consecutive. Otherwise the number of unknown lines in the force and funicular polygons will be increased. [For another method of solving this problem, see Lévy's "Statique Graphique."]

41. **Examples.** — 1. A rigid beam *AB* rests horizontally upon supports at *A* and *B*, and sustains loads as follows: Its own weight of 100 lbs. acting at its middle point; a load of 50 lbs. at *C*; a load of 60 lbs. at *D*; and a load of 80 lbs. at *E*. The successive distances between *A*, *C*, *D*, *E*, and *B* are 4 ft., 6 ft., 7 ft., and 10 ft. Required the upward pressures on the beam at the supports.

2. A rigid beam AB is hinged at A, and rests horizontally with the end B upon a smooth horizontal surface. The beam sustains loads as in Example 1, and an additional force of 40 lbs. is applied at the middle point at an angle of 45° with the bar in an upward direction. Required the pressures upon the beam at A and B. [The pressure at A may have any direction, while the pressure at B must be vertically upward, *i.e.*, at right angles to the supporting surface. Hence, this is a particular case of Problem II.]

3. Let the end B of the bar rest against a smooth surface making an angle of 30° with the horizontal; the remaining data being as in Example 2.

4. A rigid bar 2 ft. long weighs 10 lbs., its center of gravity being 8 inches from one end. The bar rests inside a smooth hemispherical bowl of 15 inches radius. What weight must be applied at the middle point in order that the bar may rest when making an angle of 15° with the horizontal? Also, what are the reactions at the ends?

5. A uniform bar 20 inches long, weighing 10 lbs., rests with one end against a smooth vertical wall and the other end overhanging a smooth peg 10 inches from the wall. A weight P is suspended from the end so that the bar is in equilibrium when making an angle of 30° with the horizontal. Find P, and the pressures exerted on the bar by the wall and peg.

42. **Special Methods.** — Certain problems can be treated more simply by other methods than by the general method of constructing the force and funicular polygons. This is sometimes true of the following:

Problem. — A rigid body is held in equilibrium by four forces acting in known lines, only one being known in magnitude and direction. It is required to completely determine the three remaining forces. (See Fig. 15.)

Let the four forces have lines of action ab, bc, cd, da, and let

EQUILIBRIUM OF NON-CONCURRENT FORCES.   27

AB be drawn representing the magnitude and direction of the known force. Now the resultant of the forces whose lines of action are da and ab must act in a line passing through M, the point of intersection of these lines; and the resultant of the other two forces must act in a line passing through N, the point of intersection of bc and cd. But the two

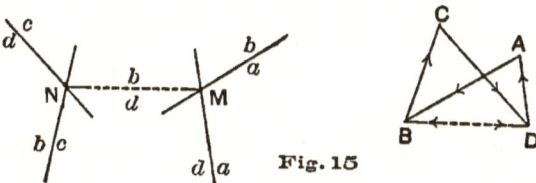

Fig. 15

resultants must be equal and opposite and have the same line of action, else there could not be equilibrium. (Art. 36.) Hence, each must act in the line MN. Draw BD parallel to MN, and AD parallel to ad; the point D being determined by their intersection. Then DA represents in magnitude and direction the force acting in da, and DB the resultant of DA and AB. But BD must represent the resultant of the two remaining forces; hence these two forces are represented by BC and CD drawn from B and D parallel respectively to bc and cd.

This problem is a special case of that treated in Art. 40. But the construction here given will in many cases be the simpler one.

Example.— A rigid body has the form of a square $ABCD$, the side AB being horizontal, and BC vertical. The weight of the body is 100 lbs., its center of gravity being at the intersection of the diagonals. It is held in equilibrium by three forces as follows: P_1 acting at C in the line AC; P_2 acting at D in the line AD; and P_3 applied at B and acting in the line joining B with the middle point of AD. Required to completely determine P_1, P_2, and P_3.

§ 3. *Resolution into Non-concurrent Systems.*

43. To Replace a Force by Two Non-concurrent Forces. — This may be done in an infinite number of ways. The lines of action of the two components must intersect at a point on the line of action of the given force, and they must further satisfy the same conditions as concurrent forces. (Art. 21.)

44. To Replace a Force by More Than Two Non-concurrent Forces. — This may be done by first resolving the given force into two forces by the preceding article, and then resolving one or both of its components in the same way. This problem and that of Art. 43 are indeterminate. (See note, Art. 20.) To make such a problem determinate, something must be specified regarding the magnitudes and lines of action of the required components. We shall consider some of the particular cases which are of frequent use in the treatment of practical problems.

45. Resolution of a Force into Two Parallel Components. — Let it be required to resolve a force into two components having given lines of action parallel to its own.

If the given force be reversed in direction, it will form with the required components a system in equilibrium. The components may then be determined by the method of Art. 38.

Example. — Let the student solve two particular cases of this problem, taking the line of action of the given force (1) between those of the components, (2) outside those of the components.

46. Resolution of a Force into Three Components. — *Problem.* — To resolve a force into three components having known lines of action.

If the given force be reversed in direction, it will form, with the required forces, a system in equilibrium. Hence these forces may be determined by either of the two methods given

RESOLUTION INTO NON-CONCURRENT SYSTEMS. 29

in Arts. 40 and 42. Or, the following reasoning may be employed, leading to the same construction as that of Art. 42.

Let AD (Fig. 16) represent the given force in magnitude and direction, and ad its line of action; the lines of action of the components being given as ab, bc, cd. Since the given force may be assumed to act at any point in the line ad, let its point of application be taken at M, the point of intersection of

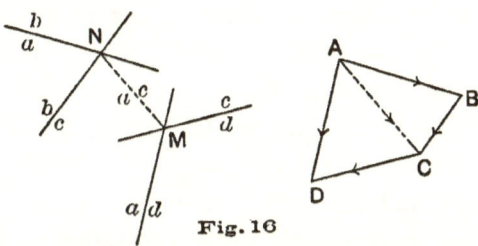

Fig. 16

ad and cd. Resolve it into two components acting in the lines cd and MN, N being the point of intersection of ab and bc. These components are represented in magnitude and direction by AC, CD. Let AC, acting in the line MN (also marked ac), be resolved into components having lines of action ab, bc. These components are given in magnitude and direction by AB, BC, drawn parallel respectively to ab, bc. Hence, the given force, acting in ad, is equivalent to the three forces represented in magnitude and direction by AB, BC, CD, acting in the lines ab, bc, cd.

If the line of action of the given force does not intersect any one of the given lines ab, bc, cd, within the limits of the drawing, it may be replaced by two components; then each may be resolved in accordance with the above method, and the results combined. If the three lines ab, bc, cd, are all parallel to ad, or if the four lines intersect in a point, the problem is indeterminate. This is evident from the preceding articles, since, by methods already given, a part of AD can be replaced by two components acting in any two of the given lines, as ab, bc; and the remaining part by two components acting in ab, cd;

or in *bc*, *cd*. This construction evidently admits of infinite variation.

For another method of solving the above problem, see Clarke's "Graphic Statics," p. 16.

§ 4. *Moments of Forces and of Couples.*

47. **Moment of a Force.** — *Definition.* — The moment of a force with respect to a point is the product of the magnitude of the force into the perpendicular distance of its line of action from the given point. The moment of a force with respect to an axis perpendicular to the force is the product of the magnitude of the force by the perpendicular distance from the axis to the line of action of the force.

If the moment is taken with respect to a point, that point is called the *origin of moments*. The perpendicular distance from the origin, or axis, to the line of action of the force is called the *arm*.

Rotation tendency of a force. — The moment of a force measures its tendency to produce rotation about the origin, or axis. Thus, if a rigid body is fixed at a point, but free to turn about that point in a given plane, any force acting upon it in that plane will tend to cause it to rotate about the fixed point. The amount of this tendency will be proportional both to the magnitude of the force and to the distance of its line of action from the given point; that is, to the moment of the force with respect to the point, as above defined.

Rotation in any plane may have either of two opposite directions, which may be distinguished from each other by signs plus and minus. Rotation with the hands of a watch supposed placed face upward in the plane of the paper will be called negative, and the opposite kind positive. It would be equally legitimate to adopt the opposite convention, but the method here adopted agrees with the usage of the majority of writers.

The sign of the moment of a force is regarded as the same as that of the rotation it tends to produce about the origin.

Moment represented by the area of a triangle. — If a triangle be constructed having for its vertex the origin of moments, and for its base a length in the line of action of a force, numerically equal to its magnitude, then the moment of the force is numerically equal to double the area of the triangle. This follows at once from the definition of moment.

48. Moment of Resultant of Two Non-parallel Forces. — *Proposition.* — The moment of the resultant of two non-parallel forces with reference to a point in their plane is equal to the algebraic sum of their separate moments with reference to the same point.

In Fig. 17, let AB, BC, and AC represent in magnitude and direction two forces and their resultant; and let ab, bc, ac be their lines of action, intersecting in a point N. Choose any point M as the origin of moments. Lay off $NP = AB$; $NQ = BC$; and $NR = AC$. Then the moments of the three forces are respectively equal to double the areas of the triangles MNP, MNQ, MNR. These three triangles have a common side MN, which may be considered the base; hence their areas are proportional to their altitudes measured from that side.

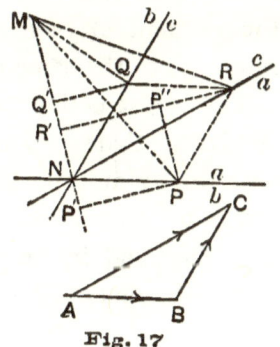

Fig. 17

These altitudes are PP', QQ', RR', perpendicular to MN. Now, if PP'' is parallel to MN, $P''R'$ is equal to PP'; also RP'' equals QQ' (since they are homologous sides of the equal triangles NQQ', PRP''). Hence $RR' = PP' + QQ'$, and therefore

Area MNR = area MNP + area MNQ;

which proves the proposition.

If the origin of moments is so taken that the moments of AB and BC have opposite signs, the demonstration needs modification. The student should attempt the proof of this case for himself.

49. Moment of a Couple. — *Definition.* — The moment of a couple about any point in its plane taken as an origin is the algebraic sum of the moments of the two forces composing it with reference to the same origin.

Proposition. — The moment of a couple is the same for every origin in its plane, and is numerically equal to the product of the magnitude of either force into the arm of the couple. (Art. 7.)

The proof of this proposition can readily be supplied by the student.

50. Moment of the Resultant of Any System. — *Proposition.* — The algebraic sum of the moments of any given complanar forces, with reference to any origin in their plane, is equal to the moment of their resultant force or resultant couple with reference to the same origin.

The construction employed in proving this proposition is similar to that used in Art. 27, which the student may profitably review at this point. Referring to Fig. 9, let the given forces be represented in magnitude and direction by AB, BC, CD, DE, and in lines of action by ab, bc, cd, dc. Let the origin of moments be any point in the space diagram. As in Art. 27, replace AB by AO, OB, acting in lines ao, ob; replace BC by BO, OC, acting in lines bo, oc; replace CD by CO, OD, acting in lines co, od; and replace DE by DO, OE, acting in lines do, oe. (For brevity, we refer to a force as AB, meaning "the force represented in magnitude and direction by AB.")

Now by Art. 48, we have, whatever the origin,

Moment of AB = moment of AO + moment of OB;
"   " BC =   "   " BO +   "   " OC;
"   " CD =   "   " CO +   "   " OD;
"   " DE =   "   " DO +   "   " OE.

Since the forces represented by BO and OB have the same line of action, their moments are numerically equal but have opposite signs; and similar statements are true of CO and OC, and of DO and OD. Hence, the addition of the above four equations shows that the sum of the moments of AB, BC, CD, DE, is equal to the sum of the moments of AO and OE. Now the given system has either a resultant force or a resultant couple. In the first case the resultant of the system is the resultant of AO and OE, and its moment is equal to the algebraic sum of their moments, by Art. 48. In the second case (which occurs only when E coincides with A), the resultant couple is composed of AO and OE (which in this case are equal and opposite forces), and the moment of the couple is, by definition, equal to the algebraic sum of the moments of AO and OE. Hence, in either case, the proposition is true.

It should be noticed that the proof here given applies to systems of parallel forces, as well as to non-parallel systems. The proposition of Art. 48 could be extended to the case of any number of forces, by considering first the resultant of two forces, then combining this resultant with the third force, and so on; but the method would fail if, in the process, it became necessary to combine parallel forces. The method here adopted is not subject to this failure.

51. **Condition of Equilibrium.** — It follows from what has preceded, that *if a given system is in equilibrium, the algebraic sum of the moments must be zero, whatever the origin.* For, in case of equilibrium, AO and OE (Fig. 9) are equal and opposite and have the same line of action; hence, the sum of their moments (which is the same as the sum of the moments of the given forces) is equal to zero. Conversely, *If the algebraic sum of the moments is zero for every origin, the system must be in equilibrium.* For, if it is not, there is either a resultant force or a resultant couple. But the moment of a force is not zero for any origin not on its line of action; and the moment of a couple is not zero for any

origin. For a fuller discussion of the conditions of equilibrium, see Arts. 58 and 59.

52. Equivalent Couples. — *Proposition.* — If a system has for its resultant a couple, it is equivalent to any couple whose moment is equal to the sum of the moments of the forces of the system.

For, as already seen (Art. 31), when the resultant is a couple, the force polygon is closed. Let the initial and final points of the force polygon coincide at some point A, and let O be the pole. Then the forces of the resultant couple are represented in magnitude and direction by AO, OA. Since the position of O is arbitrary, the force AO (or OA) may be made anything whatever in magnitude and direction. Also the line of action of the force AO may be taken so as to pass through any chosen point. Hence, the resultant couple may have for one of its forces any force whatever in the plane of the given system; and the other force will have such a line of action that the moment of the couple will be equal to the sum of the moments of the given forces.

This reasoning is equally true if the given system is a couple. Hence, a couple is equivalent to any other couple having the same moment. In other words, *all couples whose moments are equal are equivalent;* and conversely, *all equivalent couples have equal moments.*

[NOTE. — The construction above discussed fails if all forces of the given system are parallel to the direction chosen for the forces of the resultant couple. For then the force polygon is a straight line, and if the pole is chosen in that line, the strings of the funicular polygon are parallel to the lines of action of the forces, and the polygon cannot be drawn. But in this case, the system may first be reduced to a couple whose forces have some other direction, and this couple may be reduced to one whose forces have the direction first chosen. Hence, the proposition stated holds in all cases.]

53. Moment of a System. — *Definition.* — The moment of a system of forces is the algebraic sum of the moments of the forces of the system.

GRAPHIC DETERMINATION OF MOMENTS.   35

54. Moments of Equivalent Systems. — *Proposition.* — The moments of any two equivalent systems of complanar forces with respect to the same origin are equal.

This follows immediately from the preceding articles. For, if the two systems are equivalent, each is equivalent to the same resultant force or resultant couple, and the moments of the two systems are therefore each equal to the moment of this resultant and hence to each other.

§ 5. *Graphic Determination of Moments.*

55. Proposition. — If, through any point in the space diagram a line be drawn parallel to a given force, the distance intercepted upon it by the two strings corresponding to that force, multiplied by the pole distance of the force, is equal to the moment of the force with respect to the given point.

By the strings "corresponding to" a given force are meant the two strings which intersect at a point on its line of action.

Let AB (Fig. 18) represent the magnitude and direction of a force whose line of action is ab, and let M be the origin of moments. Let O be the pole, and OK ($=H$) the pole distance of the given force. Draw the strings oa, ob, and through M draw a line parallel to ab, intersecting oa and ob in P and Q.

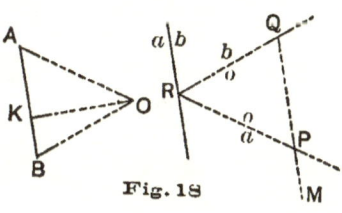

Fig. 18

Then it is to be proved that the moment of the given force with respect to M is equal to $H \times PQ$.

Let h equal the perpendicular distance of M from ab. Then the required moment is $AB \times h$. But since the similar triangles OAB and RPQ have bases AB, PQ, and altitudes H, h, respectively, it follows that

$$\frac{AB}{PQ} = \frac{H}{h}; \text{ hence, } AB \times h = PQ \times H,$$

which proves the proposition.

It should be noticed that PQ represents a *length*, while H represents a *force magnitude*. Hence, the moment of the given force with respect to M is equal to the moment of a force H with an arm PQ. [It may, in fact, be shown that the given force is equivalent to an equal force acting in the line PQ (whose moment about M is therefore zero), and a couple with forces of magnitude H, and arm PQ. For AB acting in ab is equivalent to AO and OB acting in ao and ob respectively. Also, AO acting in ao is equivalent to forces represented by AK and KO acting respectively in PQ and in a line through P at right angles to PQ; and OB may be replaced by forces represented by OK and KB, the former acting in a line through Q perpendicular to PQ, and the latter in PQ. But AK and KB are equivalent to AB; hence, the proposition is proved.]

56. Moment of the Resultant of Several Forces. — The moment of the resultant of any number of consecutive forces in the force and funicular polygons may be found by a method similar to that just described. Thus, let Fig. 19 represent the force polygon and the funicular polygon for six forces, and let it be required to find the moment of the resultant of the four forces represented in the force polygon by BC, CD, DE, and EF, with respect to any point M. The resultant of the four forces is represented by BF, and acts in a line through the intersection of ob and of. Through M draw a line parallel to BF, intersecting ob and of in P and Q respectively. Then PQ multiplied by OK, the pole distance of BF, gives the required moment. This method does not apply to the determination of the moment of the resultant of several forces *not* consecutive in the force polygon.

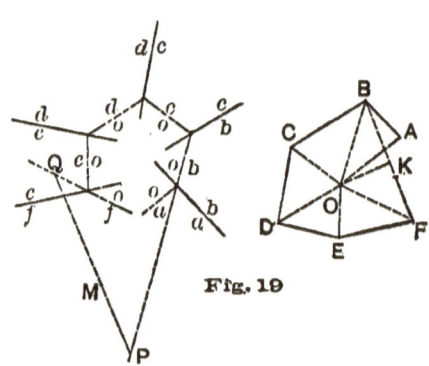

Fig. 19

57. Moments of Parallel Forces. — The method of Arts. 55 and 56 is especially useful when it is desired to find the moments of any or all of a system of parallel forces; since with such a system the pole distance is the same for all forces, and the moments are therefore proportional to the intercepts found by the above method.

Example. — Assume five parallel forces at random; choose an origin, and determine their separate moments, also the moment of their resultant, by the method of Arts. 55 and 56.

§ 6. *Summary of Conditions of Equilibrium.*

58. Graphical and Analytical Conditions of Equilibrium Compared. — It has been shown (Art. 35) that the conditions of equilibrium for a system of complanar forces acting on a rigid body are two in number:
(I) The force polygon must close.
(II) Any funicular polygon must close.
The analytical conditions * are the following:
(1) The algebraic sum of the resolved parts of the forces in any direction must be zero.
(2) The algebraic sum of their moments for any origin must be zero.
The condition (1) is readily seen to be equivalent to (I). For if the sides of the force polygon be orthographically projected upon any line, their projections will represent in magnitude and direction the resolved parts of the several forces parallel to the line; and, if the force polygon is closed, the algebraic sum of these projections is zero, whatever the direction of the assumed line. (See Art. 22.) It may also be seen that condition (II) carries with it (2). For, if every funicular polygon closes, the system is equivalent to two equal and opposite forces having the same line of action (Arts. 31,

---

* See *Minchin's Statics*, Vol. I, p. 114.

35, and 50); and the sum of the moments of these two forces must be zero.

A further comparison may be made. The analytical condition (2) carries (1) with it; and similarly the graphical condition (II) carries with it the condition (I). That (2) includes (1) may be seen as follows: If the sum of the moments is zero for one origin M_1, there can be no resultant couple, neither can there be a resultant force unless with a line of action passing through M_1. If the sum of the moments is zero for a second origin M_2, the resultant force, if one exists, must act in the line M_1M_2. If the sum of the moments is zero also for a third origin M_3, not on the line M_1M_2, there can be no resultant force. It follows at once that condition (1) must hold.

That (II) includes (I) may be shown as follows: Let A and E be the first and last points of the force polygon, and choose a pole O_1. Then, if the funicular polygon closes, O_1A and O_1E are parallel. Choose a second pole O_2, and, if the funicular polygon again closes, O_2A and O_2E are parallel. AE must then be parallel to O_1O_2, unless A and E coincide. Now, choose a third pole O_3, not on O_1O_2; if the funicular polygon for this pole closes, O_3A and O_3E must be parallel. But this is impossible unless A and E coincide; that is, unless condition (I) holds.

The last result may be reached in another way. With any pole O draw a funicular polygon and suppose it to close. The system is thus reduced to two forces acting in the same line oa. Hence, there is no resultant couple, and if there is a resultant force, its line of action is oa. With the same pole draw a second funicular polygon, the first side being $o'a'$, parallel to oa. If this polygon closes, there can be no resultant force, for if one existed it would act in the line $o'a'$; and there would thus be two resultant forces, acting in different lines oa and $o'a'$, which is impossible.

SUMMARY OF CONDITIONS OF EQUILIBRIUM.

59. Summary. — It is now evident that the conditions necessary to *insure* equilibrium may be stated in several different ways, both analytically and graphically. To summarize :

A. Analytically : There will be equilibrium if either of the following conditions is satisfied :

(1) The sum of the moments is zero for each of three points not in the same line.

(2) The sum of the moments is zero for each of two points, and the sum of the resolved parts is zero in a direction not perpendicular to the line joining those two points.

(3) The sum of the moments is zero for one point, and the sum of the resolved parts is zero for each of two directions.

B. Graphically : There will be equilibrium if either of these three conditions is satisfied :

(1) A funicular polygon closes for each of three poles not in the same line.

(2) Two funicular polygons close for the same pole.

(3) One funicular polygon closes and the force polygon closes.

CHAPTER III. INTERNAL FORCES AND STRESSES.

§ 1. *External and Internal Forces.*

60. **Definitions.** — It was stated in Art. 3 that every force acting upon any body is exerted by some other body. In what precedes, we have been concerned only with the effects produced by forces upon the bodies to which they are applied. It has therefore not been needful to consider the bodies which exert the forces. It is now necessary to consider forces in another aspect.

The forces applied to any particle of a body may be either *external* or *internal*.

An *external force* is one exerted upon the body in question by some other body.

An *internal force* is one exerted upon one portion of the body *by another portion of the same body*.

It is important to note, however, that the same force may be internal from one point of view, and external from another. Thus, if a given body be conceived as made up of two parts, X and Y, a force exerted *upon X by Y* is internal as regards the whole body, but external as regards the part X. Thus, let

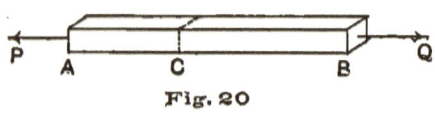
Fig. 20

AB (Fig. 20) represent a bar, acted upon by two forces of equal magnitude applied at the ends parallel to the length of the bar, in such a way as to tend to pull it apart. These two forces are exerted upon AB by some other bodies not specified. If the whole bar be considered, the external forces acting upon it are simply the two forces named.

But suppose the body under consideration is AC, a portion of AB. The external forces acting upon this body are (1) a force at A, already mentioned, and (2) a force at C, exerted upon AC by CB. This latter force is *internal* to the bar AB, but external to AC.

61. Conditions of Equilibrium Apply to External Forces. — In applying the conditions of equilibrium deduced in previous articles, it must be remembered that only external forces are referred to. It is also important to notice that the principles apply to *any body* or *any portion of a body* in equilibrium; and the system of forces in every case must include all forces that are external to the body or portion of a body in question.

Thus, if the bar AB (Fig. 20) be in equilibrium under the action of two opposite forces P and Q, applied at A and B respectively, as shown in the figure, the principles of equilibrium may be applied either to the whole bar, or to any part of it, as AC.

(*a*) For the equilibrium of the whole bar AB, we must have $P=Q$, these being supposed the only *external* forces acting on the bar.

(*b*) For the equilibrium of AC, the force exerted upon AC by CB must be equal and opposite to P. This latter force is *external* to AC, though *internal* to the whole bar.

The method just illustrated is of frequent use in the investigation of engineering structures. It is often desired to determine the internal forces acting in the members of a structure, and the general method is this: Direct the attention to such a portion of the whole structure or body considered that the internal forces which it is desired to determine shall be external to the portion in question. (See Art. 67.)

§ 2. *External and Internal Stresses.*

62. Newton's Third Law. — Let X and Y be any two portions of matter; then if X acts upon Y with a certain force, Y acts upon X with a force of equal magnitude in the opposite direction. This is the principle stated in Newton's third law of motion, — that "to every action there is an equal and contrary reaction." It is justified by universal experience.

63. Stress. — *Definition.* — Two forces exerted by two portions of matter upon each other in such a way as to constitute an action and its reaction, make up a *stress*.

Illustrations. — The earth attracts the moon with a certain force, and the moon attracts the earth with an equal and opposite force. The two forces constitute a stress.

Two electrified bodies attract (or repel) each other with equal and opposite forces. These two forces constitute a stress.

Any two bodies in contact exert upon each other equal and opposite pressures (forces), constituting a stress.

By the *magnitude of a stress* is meant the magnitude of either of its forces.

64. External and Internal Stresses. — It has been seen that two portions of matter are concerned in every stress. Now the two portions may be regarded either as separate bodies, or as parts of a body or system of bodies which includes both.

A stress acting between two parts of the same body (or system of bodies) is an *internal stress* as regards that body or system.

A stress acting between two distinct bodies is an *external stress* as regards either body.

It is important to notice that the same stress may be internal from one point of view, and external from another. Thus, if a given body be considered as made up of two parts X and Y, a stress exerted between X and Y is *internal* to the whole body, but external to either X or Y.

Illustration. — Consider a body AB (Fig. 21) resting upon a second body Y, and supporting another body X, as shown. If the weight of the body AB be disregarded, the forces acting upon it are (1) the downward pressure (say P) exerted by X at the surface A, and the upward pressure (say Q) exerted by Y at B; these forces being equal and opposite, since the body is in equilibrium. Now the body X is acted upon by AB with a force equal and opposite to P, and these two forces constitute a stress which is external to AB. There is also an external stress exerted between AB and Y at B. But let AB be considered as made up of two parts, AC and CB. Then (Art. 60) CB exerts upon AC a force upward, and AC exerts upon CB a force downward. These two forces are an action and its reaction, and constitute a stress which is internal to the body AB. This same stress is, however, external to either AC or CB. An equivalent stress evidently exists at every section between A and B. (When we refer to the force acting upon CB at C, we mean the *resultant* of all forces exerted upon the particles of CB by the particles of AC. This resultant is made up of very many forces acting between the particles. Also the stress at C means the stress made up of the two resultants of the forces exerted by AC and CB upon each other.)

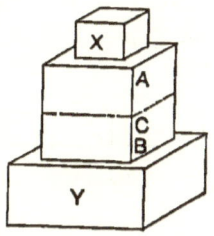

Fig. 21

65. Three Kinds of Internal Stress. — It is evident that the internal stress at C in the body AB (Fig. 20) depends upon the external forces applied to the body. If the forces at A and B cease to act, the forces exerted by AC and CB upon each other become zero. If the forces at A and B are reversed in direction, so also are those at C. (As a matter of fact, the particles of AC exert forces upon those of CB, even if the external forces do not act. But if the external forces applied to CB are balanced, the resultant of the forces exerted on CB by AC is zero.)

The nature of the internal stress at any point in a body is thus seen to depend upon the external forces applied to the body.

Now, if we consider two adjacent portions of a body (as the parts X and Y, Fig. 22) separated by a plane surface, the external forces may have either of three tendencies: (1) to pull X and Y apart in a direction perpendicular to the plane of separation; (2) to push them together in a similar direction; (3) to slide each over the other along the plane of separation. Corresponding to these three tendencies, the stress between X and Y may be of either of three kinds: *tensile, compressive,* or *shearing.*

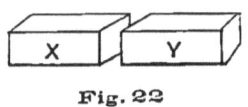

Fig. 22

A *tensile stress* is such as comes into action to resist a tendency of the two portions of the body to be pulled apart in the direction of the normal to their surface of separation.

A *compressive stress* is such as comes into action to resist a tendency of X and Y to move toward each other along the normal to the surface.

A *shearing stress* is such as acts to resist a tendency of X and Y to slide over each other along the surface between them.

In case of a tensile stress, the force exerted by X upon Y has the direction from Y toward X; and the force exerted by Y upon X has the direction from X toward Y.

In case of a compressive stress, the force exerted by X upon Y has the direction from X toward Y; and the force exerted by Y upon X has the direction from Y toward X.

In the case of a shearing stress, the force exerted by X upon Y may have any direction in the plane of separation; the force exerted by Y upon X having the opposite direction.

If X and Y are separate bodies, instead of parts of one body, a similar classification may be made of the kinds of stress between them; but with these we shall have no occasion to deal. The terms *tensile stress, compressive stress,* and *shearing stress* (or *tension, compression,* and *shear*) are usually applied only to internal stresses.

EXTERNAL AND INTERNAL STRESSES. 45

66. Strain. — In what has preceded, the bodies dealt with have been regarded as *rigid;* that is, the relative positions of the particles of any body have been regarded as remaining unchanged. But, as remarked heretofore, no known body is perfectly rigid. If no external forces act upon a body, its particles take certain positions relative to each other, and the body has what is called its *natural* shape and size. If external forces are applied, the shape and size will generally be changed; the body is then said to be in a *state of strain*. The deformation produced by any system of applied forces is called the *strain* due to those forces. The nature of this strain in any case depends upon the way in which the forces are applied. It is unnecessary to treat this subject further at this point, since we shall at present be concerned only with problems in the treatment of which it will be sufficiently correct to regard the bodies as rigid.

[NOTE. — There is a lack of uniformity among writers in regard to the meanings attached to the words *stress* and *strain*. It may, therefore, be well to explain again at this point the way in which these words are used in the following pages. The word *stress* should be employed only in the sense above defined, as consisting of two equal and opposite forces constituting an action and its reaction. The two forces are exerted respectively *by* two bodies or portions of matter *upon* each other. An *internal stress* is a stress between two parts of the same body. An *internal force* is one of the forces of an internal stress. It is intended in what follows to use the words "internal stress" (or simply "stress") only when *both* the constituent forces are referred to; and when only one of the forces is meant, to use the words "internal force" (or simply "force"). It will be noticed, therefore, that in the following pages the words "*force* in a bar" are frequently used where many writers would say "*stress*." This departure from the usage of many high authorities seems justified by the following considerations: (1) It agrees with the usage which is being adopted by the highest authorities in pure mechanics. (2) It is desirable that the nomenclature of technical mechanics shall agree with that of pure mechanics, so far as they deal with the same conceptions. The definition of strain above given is in conformity with the usage of the majority of the more recent text-books. But it is not rare to find in technical literature the word *strain* used in the sense of *internal stress* as above defined. Such use of the word should be avoided.]

§ 3. Determination of Internal Stresses.

67. General Method. — The stresses exerted between the parts of a body may or may not be completely determinate by means of the principles already deduced. But in all cases these principles suffice for their partial determination. The general method employed is always the same, and will now be illustrated. As heretofore we deal only with complanar forces.

Let XY (Fig. 23) represent a body in equilibrium under the action of any known external forces as shown. Now conceive the body to be divided into two parts as X and Y, separated by any surface. The particles of X near the surface exert upon those of Y certain forces, and are in return acted upon by forces exerted by the particles of Y. These forces are internal as regards the whole body. In order to determine them so far as possible, we proceed as follows: Let the resultant of all the forces exerted by Y upon X be called T; then T is either a single force or a couple. Now apply the conditions of equilibrium to the body X. The external forces acting on X are P_1, P_2, P_3, and T. Since P_1, P_2, and P_3 are supposed known, T can be determined. In fact, T is equal and opposite to the resultant of P_1, P_2, and P_3.

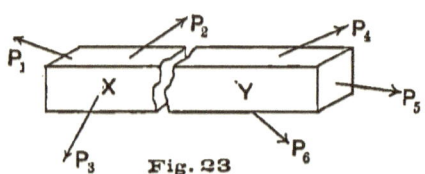

Fig. 23

So much can always be determined. But T is the resultant of a great number of forces acting on the various particles of X; and these separate forces cannot in general be determined by methods which lie within the scope of this work.

The general principle just illustrated may be stated as follows:

If a body in equilibrium under any external forces be conceived as made up of two parts X and Y, then the internal forces exerted by X upon Y, together with the external forces acting on Y, form a system in equilibrium.

DETERMINATION OF INTERNAL STRESSES.

As an immediate consequence, we may state that *the resultant of the forces exerted by X upon Y is equivalent to the resultant of the external forces acting on X; and is equal and opposite to the resultant of the external forces acting upon Y.*

Example. — Assume a bar of known dimensions, and the magnitudes, directions, and points of application of five forces acting on it. Then (1) determine a sixth force which will produce equilibrium; and (2) assume the bar divided into two parts and find the resultant of the forces exerted by each part on the other.

68. Jointed Frame. — In certain ideal cases (corresponding more or less closely to actual cases), the internal forces may be more completely determined. The most important of these cases is that which will be now considered. Conceive a rigid body made up of straight rigid bars hinged together at the ends. Assume the following conditions:

(1) The hinges are without friction.

(2) All external forces acting on the body are applied at points where the bars are joined together.

The meaning of these conditions will be seen by reference to Fig. 24. The three bars X, Y, and Z are connected by a "pin joint," the end of each bar having a hole or "eye" into which is fitted a pin. (Of course the three bars cannot be in the same plane, but they may be nearly so, and will be so assumed in what follows.) Condition (1) is satisfied if the pin is assumed frictionless. The effect of this is that the force exerted upon the pin by any bar (and the equal and opposite reaction exerted upon the bar by the pin) act in the normal to the surfaces of these bodies at the point of contact; and, therefore, through the centers of the pin and the hole. Condition (2) means that any external force (that is, external to the

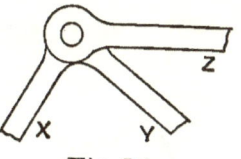

Fig. 24.

whole body) applied to any bar is applied to the end and in a line through the center of the pin.

With the connection as shown, the bars do not exert forces upon each other directly. But each exerts a force upon the pin, and any force exerted by Y or Z upon the pin causes an equal force to be exerted upon X. (This is seen by applying the condition of equilibrium to the pin.) Hence, in considering the forces acting upon any bar as X, we may disregard the pin and assume that each of the other bars acts directly upon X. By what has been said, all such forces exerted upon X by the other bars meeting it at the joint may be regarded as acting at the same point — the center of the pin. We therefore treat the bars as mere "material lines," and regard all forces exerted on any bar (whether by the other bars or by outside bodies) as applied at the ends of this "material line."

Since, with these assumptions, all forces acting on any bar MN (Fig. 25) are applied either at M or N, the forces applied at M must balance those applied at N. The resultants of the two sets must therefore be equal and opposite, and have the same line of action — namely MN. Further, it follows that the stress in the bar, acting across any plane perpendicular to its length, is a direct tension or compression.

69. **Internal Stresses in a Jointed Frame.** — Let Fig. 25 represent a jointed frame such as above described, in equilibrium under any known external forces. Let us apply the general method of Art. 67 to this case. Divide the body into two parts, X and Y, by the surface AB as shown. Now apply the conditions of equilibrium to the body X. The system of forces acting upon this body consists of P_1, P_2, P_6, and the forces exerted by Y upon X in the three members cut by the surface AB. By Art. 68 the lines of action of these forces are known,

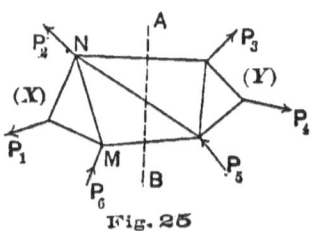

Fig. 25

DETERMINATION OF INTERNAL STRESSES.

being coincident with the axes of the members cut. Hence, the system in equilibrium consists of six forces, three completely known, and three known only in lines of action. The determination of the unknown forces in magnitude and direction is then a case under the general problem discussed in Art. 40.

Nature of the stresses. — As soon as the direction of the force acting upon X in any one of the members cut is known, the nature of the stress in that member (whether tension or compression) is known. For a force *toward* X denotes compression; while a force *away from* X denotes tension. (Art. 65.)

If a section can be taken cutting only two members, the forces in these may be found by the force polygon alone. The same is true, if any number of members are cut, but the stresses in all but two are known.

The methods described in the last three articles will find frequent application in the chapters on roof and bridge trusses, Part II.

Example. — Assume a jointed frame similar to the one shown in Fig. 25, and let external forces act at all the joints. Then (1) assume all but three of the forces known in magnitude and direction and determine the remaining three so as to produce equilibrium. (2) Take a section cutting three members and determine the stresses in those members.

70. **Indeterminate Cases.** — If, in dividing the frame, more than three members are cut, the number of unknown forces is too great to admit of the determination of their magnitudes. In such a case, it may happen that a section elsewhere through the body will cut but three members; and that after the determination of the stresses in these three, another section can be taken cutting but three members whose stresses are unknown. So long as this can be continued, the determination of the

GRAPHIC STATICS.

internal stresses can proceed. Thus, in Fig. 26, if a section be first taken at AB, there are four unknown forces to be determined. But, if the section $A'B'$ be first taken, the stresses in the three members cut may be determined; after which the section AB will introduce but three unknown stresses.

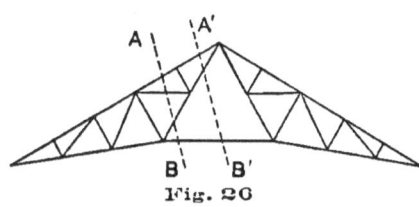

Fig. 26

There may, however, be cases in which the stresses cannot all be determined by any method. With such actually indeterminate cases we shall not usually have to deal. It should be noticed, also, that even when only three members are cut, the problem is indeterminate *if these three intersect in a point*. As in the case just discussed, this indeterminateness may be either actual or only apparent; in the latter case it may be treated as above indicated.

No attempt is here made to develop all methods that are applicable or useful in the determination of stresses in jointed frames. Some of these are best explained in connection with the actual problems giving rise to them. We have sought here only to explain and clearly illustrate *general principles*.

71. Funicular Polygon Considered as Jointed Frame. — Let ab, bc, cd, da (Fig. 27) be the lines of action of four forces in equilibrium, the force polygon being $ABCDA$. Choosing a

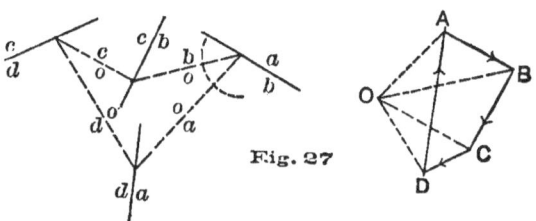

Fig. 27

pole, draw any funicular polygon, as the one shown. Now let the body upon which the forces act be replaced by a jointed

DETERMINATION OF INTERNAL STRESSES. 51

frame whose bars coincide with the sides of the funicular polygon. If at the joints of this frame the given forces be applied, the frame will be in equilibrium; and each bar will sustain a tension or compression whose magnitude is represented by the corresponding ray of the force diagram.

To prove this, we apply the "general method" of Art. 67. Consider any joint (as the intersection of oa and ob), and let the frame be divided by a plane cutting these two members. Then the portion of the frame about the joint is acted upon by three forces: AB, acting in the line ab, and forces acting in the bars cut, their lines of action being oa, ob. If the bar oa sustains a compression and ob a tension, their magnitudes being represented by OA and BO respectively, the portion of the frame about the joint will be in equilibrium. Hence, the tendency of the force AB is to produce the stresses mentioned in the bars oa, ob. In the same way it may be shown that the tendency of the force BC is to produce in ob and oc tensile stresses of magnitudes OB and CO, respectively. Applying the same reasoning to each joint, it is seen that every part of the frame will be in equilibrium if the bars sustain stresses as follows: The bar oa must sustain a compression OA; ob a tension OB; oc a tension OC; and od a compression OD. Hence, if the bars are able to sustain these stresses, the frame will be in equilibrium.

If the stress in any member of the frame is a tension, that member may be replaced by a flexible string. This is the origin of the name string as applied to the sides of the funicular polygon. This name is retained for convenience, but, as just shown, it is not always appropriate.

72. Outline of Subject.—The foregoing pages, embracing Part I, have been devoted to a development of the principles of pure statics. We pass next to the application of these principles to special classes of problems.

Part II treats of the determination of internal stresses in

engineering structures. Only "simple" structures are considered, — that is, those whose discussion does not involve the theory of elasticity. The structures considered include roof trusses, beams, and bridge trusses.

Part III develops the graphic methods of determining centroids (centers of gravity) and moments of inertia of plane areas, including a short discussion of "inertia-curves."

PART II.

STRESSES IN SIMPLE STRUCTURES.

CHAPTER IV. INTRODUCTORY.

§ 1. *Outline of Principles and Methods.*

73. The Problem of Design. — When any structure is subjected to the action of external forces, there are brought into action certain *internal stresses* in the several parts of the structure. The nature and magnitudes of these stresses depend upon the external forces acting (Art. 65). In designing the structure, each part must be so proportioned that the stresses induced in it will not become such that the material cannot sustain them without injury.

To determine these internal stresses, when the external forces are wholly or partly given, is the problem of design, so far as it will be here treated.

74. External Forces. — The external forces acting on a structure must generally be completely known before the internal stresses can be determined. These external forces are usually only partly given, and the first thing necessary is to determine them fully.

The external forces include (1) the *loads* which the structure is built to sustain, and (2) the *reactions* exerted by other bodies upon the structure at the points where it is supported. The former are known or assumed at the outset and the latter are to be determined.

75. Two Classes of Structures. — Structures may be divided into two classes, according as they may or may not be treated as rigid bodies in determining the reactions. This may be illustrated as follows:

Let a bar AB (Fig. 28) be supported in a horizontal position at two points A and B, the supports being smooth so that the pressures on the bar at A and B are vertical. Let a known load P be applied to the beam at a given point, and let it be required to determine the reactions at A and B.

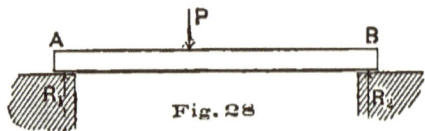
Fig. 28

This is a determinate problem; for there are three parallel forces in equilibrium, two being known only in line of action. This problem was solved in Art. 38.

But let AC (Fig. 29) be a rigid bar supported at three points, A, B, and C, the reactions at those points being vertical. Let any known loads be applied at given points, and let it be required to determine the three reactions.

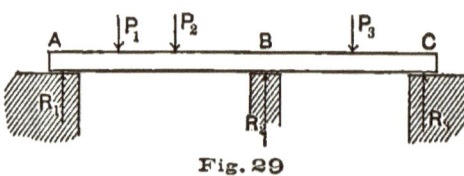
Fig. 29

This problem is indeterminate; for any number of sets of values of the three reactions may be found, which, with the applied loads, would produce equilibrium if acting on a rigid body. (See Art. 40.)

Since, however, an actual bar is not a perfectly rigid body, such a problem as the one just stated is, in reality, determinate. But it cannot be solved without making use of the elastic properties of the material of which the body is composed.

The two classes of problems are, therefore, the following: (1) those in which the reactions can be determined by treating the structure as a rigid body, and (2) those in which the determination of the reactions involves the theory of elasticity. We shall at present deal only with the former class of problems. Structures coming under this class will be called *simple* structures.

76. Truss. — A truss is a structure made up of straight bars with ends joined together in such a manner that the whole acts as a single body. The ends of the bars are, in practice, joined in various ways; but in determining the internal stresses, the connections are assumed to be such that no resistance is offered by a joint to the rotation of any member about it. Such a structure to be indeformable must be made up of triangular elements; for more than three bars hinged together in the form of a polygon cannot constitute a rigid whole. If the external forces are applied to the truss only at the points where the bars are joined, the internal stress at any section of a member will be a simple tension or compression, directed parallel to the length of the bar. (See Art. 68.)

The most important classes of trusses are roof trusses and bridge trusses. The methods used in discussing these classes are, of course, applicable to any framed structures under similar conditions.

77. Loads on a Truss. — The loads sustained by a truss may be either *fixed* or *moving*. A fixed (or *dead*) load is one whose point of application and direction remain constant. A moving (or *live*) load is one whose point of application passes through a series of positions. Fixed loads may be either *permanent* or *temporary*.

The loads on a roof truss are usually all fixed, but are of various kinds, viz., the weight of the truss itself and of the roof covering, which is a permanent load; the weight of snow lodging on the roof, and the pressure of wind, both of which are temporary loads.

A bridge truss supports both fixed and moving loads. The former include the weight of the truss itself, of the roadway, of all lateral and auxiliary bracing (permanent loads); and of snow (a temporary load). The latter consist of moving trains in the case of railway bridges, and of teams or crowds of animals or people in the case of highway bridges.

78. Combination of Stresses Due to Different Causes. — When a truss is subject to a variety of external loads, it is often convenient to consider the effect of a part of them separately. If tensile and compressive stresses are distinguished by signs plus and minus, the stress in any member due to the combined action of any number of loads is equal to the algebraic sum of the stresses due to the loads acting separately.

A proof of this proposition might be given; but it may be accepted as sufficiently evident without formal demonstration.

79. Beams. — Another class of bodies to be treated is included under the name *beam*.

A beam may be defined as a bar (usually straight) resting on supports and carrying loads. The loads and reactions are commonly applied in a direction transverse to the length of the bar; but this is not necessarily the case.

The internal stresses in any section of a beam are less simple than those in the bars of an ideal jointed frame such as a truss is assumed to be. A discussion of beams is given in Chap. VI.

80. Summary of Principles Needed. — It will be well to summarize at this point the main principles and methods which will be employed in the discussion of the problems that follow.

The general problem presented by any structure consists of two parts: (*a*) the determination of the unknown external forces (or reactions) and (*b*) the determination of the internal stresses.

(*a*) In the case of simple structures the unknown reactions are usually two in number, and the cases most commonly presented are the following:

1st. — Their lines of action are known and parallel.

2nd. — The line of action of one and the point of application of the other are known.

Since all the external forces form a system in equilibrium, these two cases fall under the general problems discussed in Arts. 38 and 39.

(*b*) In the case of a jointed frame or truss, the lines of action of all internal forces are known, since they coincide with the axes of the truss members. (Art. 68.) In determining their magnitudes we may have to deal with the following problems in equilibrium:

1st. — The system in equilibrium may be completely known, except the magnitudes of two forces.

2nd. — The magnitudes of three forces may be unknown.

The first case may be solved by simply making the force polygon close. (See Art. 35.)

The second case may be solved by the method of Art. 40, which consists in making the force and funicular polygons close; or by the method of Art. 42; or by the principle of moments (Art. 51).

The student should be thoroughly familiar with the problems and principles here referred to. In the following chapters we proceed to their application.

81. **Division of the Subject.** — The subject of the design of structures, so far as here dealt with, will be treated in three divisions. The first relates to framed structures sustaining only stationary loads; the second to beams sustaining both fixed and moving loads; the third to framed structures sustaining both fixed and moving loads. Among structures of the first class, the most important are roof trusses; hence, these are chiefly referred to in the next chapter. For a similar reason, the chapter devoted to the third class of structures refers principally to bridge trusses. The chapter on beams precedes that on bridge trusses, for the reason that the methods used in dealing with a beam under moving loads form a useful introduction to those employed in treating certain classes of truss problems.

CHAPTER V. ROOF TRUSSES.—FRAMED STRUCTURES SUSTAINING STATIONARY LOADS.

§ 1. *Loads on Roof Trusses.*

82. Weights of Trusses.—Among the loads to be sustained by a roof truss is the weight of the truss itself. Before the structure is designed, its weight is unknown. But, since it is necessary to know the weight in order that the design may be correctly made, the method of procedure must be as follows:

Make a preliminary estimate of the weight, basing it upon knowledge of similar structures; or, in the absence of such knowledge, upon the best judgment available. Then design the various truss members, compute their weight, and compare the actual weight of the truss with the assumed weight. If the difference is so great as to materially affect the design of the truss members, a new estimate of weight must be made, and the computations repeated or revised. No more than one or two such trials will usually be needed.

As a guide in making the preliminary estimate of weight, the following formulas may be used. They are taken from Merriman's "Roofs and Bridges," being intended to represent approximately the data for actual structures, as compiled by Ricker in his "Construction of Trussed Roofs."

Let $l=$ span in feet; $a=$ distance between adjacent trusses in feet; $W=$ total weight of one truss in pounds. Then for wooden trusses

$$W = \tfrac{1}{2} al \left(1 + \tfrac{1}{10} l\right);$$

and for wrought iron trusses

$$W = \tfrac{3}{4} al \left(1 + \tfrac{1}{10} l\right).$$

LOADS ON ROOF TRUSSES.

83. Weight of Roof Covering. — The weight of roof covering can be correctly estimated beforehand from the known weights of the materials. The following data may be employed, in the absence of information as to the specific material to be used. (See Merriman's "Roofs and Bridges," p. 4.) The numbers denote the weight in pounds per square foot of roof surface.

Shingling: Tin, 1 lb.; wooden shingles, 2 to 3 lbs.; iron, 1 to 3 lbs.; slate, 10 lbs.; tiles, 12 to 25 lbs.

Sheathing: Boards 1 in. thick, 3 to 5 lbs.

Rafters: 1.5 to 3 lbs.

Purlins: Wood, 1 to 3 lbs.; iron, 2 to 4 lbs.

Total roof covering, from 5 to 35 lbs. per square foot of roof surface.

84. Snow Loads. — The weight of snow that may have to be borne will differ in different localities. For different sections of the United States the following may be used as the maximum snow loads likely to come upon roofs.

Maximum for northern United States, 30 lbs. per square foot of horizontal area covered.

For latitude of New York or Chicago, 20 lbs. per square foot.

For central latitudes in the United States, 10 lbs. per square foot.

The above weights are given in Merriman's "Roofs and Bridges." They are in excess of those used by some Bridge and Roof companies.

85. Wind Pressure Loads. — The intensity of wind pressure against any surface depends upon two elements: (a) the velocity of the wind, and (b) the angle between the surface and the direction of the wind.

Theory indicates that the intensity of wind pressure upon a surface perpendicular to the direction of the wind should be proportional to the square of the velocity of the wind relative to the surface. As an approximate law this is borne out by experiment. If p denotes the pressure per unit area, and v the velocity of the wind, the law is expressed by the formula

$p = kv^2$. Here k is proportional to the density of air. Its numerical value may be taken as 0.0024, if the units of force, length, and time are the pound, foot, and second respectively.

If the wind strikes a surface obliquely, experiment shows that the resulting pressure has a direction practically normal to the surface. The tangential component is inappreciable, owing to the very slight friction between air and any fairly smooth surface. The intensity of the normal pressure depends upon the angle at which the wind meets the surface.

For a given velocity of wind let p_a denote the normal pressure per unit area, when the direction of wind makes an angle a with the surface, and p_n the pressure per unit area due to the same wind striking a surface perpendicularly. Then the following formula * has been given:

$$p_a = \frac{2 \sin a}{1 + \sin^2 a} p_n.$$

It will rarely be necessary to use values of p_n greater than 50 lbs. per square foot. The following table gives values of the coefficient of p_n in the above formula for different values of a. The value of p_n may be taken as from 40 to 50 lbs. per square foot.

a	$\dfrac{2 \sin a}{1 + \sin^2 a}$	a	$\dfrac{2 \sin a}{1 + \sin^2 a}$
0°	0.00	50°	0.97
10°	0.34	60°	0.99
20°	0.61	70°	1.00
30°	0.80	80°	1.00
40°	0.91	90°	1.00

---

* This formula is given by various writers. It is cited by Langley (" Experiments in Aerodynamics," p. 24), who attributes it to Duchemin. Professor Langley's elaborate experiments show so close an agreement with the formula that it may be used without hesitation in estimating the pressure on roofs.

§ 2. *Roof Truss with Vertical Loads.*

86. Notation. — The method of determining internal stresses in the case of vertical loading will be explained by reference to the form of truss shown in Fig. 30. The method will be seen to be independent of the particular form of the truss.

For designating the truss members and the lines of action of external forces a notation will be employed similar to that used in previous chapters. Let each of the areas in the truss diagram be marked with a letter or other symbol as shown in Fig. 30; then the truss member or force-line separating any two areas may be designated by the two symbols belonging to those areas. Thus, the lines of action of the external forces are *ab*, *bc*, *cd*, etc., and the truss members are *gh*, *hb*, *hi*, etc. It is to be noticed that the lines representing the truss members represent also the lines of action of forces, — namely, the internal forces in the members. The joint, or point at which several members meet, may be designated by naming all the surrounding letters. Thus, *bcih*, *hijg* are two such points.

87. Loads and Reactions. — The loads now considered are assumed to be applied in a vertical direction, and to act at the upper joints of the truss. This assumption as to the points of application may in some cases represent very nearly the facts; in other cases the loads will, in reality, be applied partly at intermediate points on the truss members. If the latter is the case, the load borne upon any member is assumed to be divided between the two joints at its ends. In this case the member will be subject not only to direct tension or compression, but to bending. With the latter we are not here concerned, although it must always be considered in designing the member.

The ends of the truss are supposed to be supported on horizontal surfaces, and the reaction at each point of support is assumed to have a vertical direction.

If the loading is symmetrical with reference to a vertical line through the middle of the truss, it is evident that each reaction is equal to half the total load. If the loading is not symmetrical, the reactions cannot be determined so simply. They may, however, be readily computed by either graphic or algebraic methods. Graphically, the problem is identical with that solved in Art. 38. The truss is treated as a rigid body, the external forces acting upon it being the loads and reactions, which form a system of parallel forces in equilibrium. Two of these forces (the reactions) are unknown in magnitude, but known in line of action. The construction for determining their magnitudes is as follows:

Draw the force polygon *ABCDEF* for the five loads; choose a pole *O*, draw rays *OA*, *OB*, *OC*, etc., and draw the funicular polygon as shown in Fig. 30. The two polygons are to be completed by including the reactions *FG*, *GA*, and both polygons must close. We may draw first *og*, the closing line of the funicular polygon, and then the ray *OG* parallel to it, thus determining the point *G* in the force polygon. The reactions are now shown in magnitude and direction by *FG* and *GA*.

88. **Determination of Internal Stresses.** — When the external forces are all known, the internal stresses may be found very readily. The only principle needed is, that for any system of forces in equilibrium, the force polygon must close. The construction will now be explained.

Considering any joint of the truss (Fig. 30) as *ghijg*, fix the attention upon the portion of the truss bounded by the broken line in the figure. This portion is a body in equilibrium under the action of four forces whose lines of action coincide with the axes of the four bars *gh*, *hi*, *ij*, *jg*, respectively. These forces are *internal* as regards the truss as a whole, but *external* to the part in question; each force being one of the pair constituting the internal stress at any point of the bar. Such a force acts *from* the joint if the stress in the bar is a tension; *toward* it if the stress is a compression. (Art. 69.)

ROOF TRUSS WITH VERTICAL LOADS.

Since these four forces form a system in equilibrium, their force polygon must close. This condition will enable us to fully determine the magnitudes of the forces, provided all but two are known, since the polygon can then be constructed as in Art. 18.

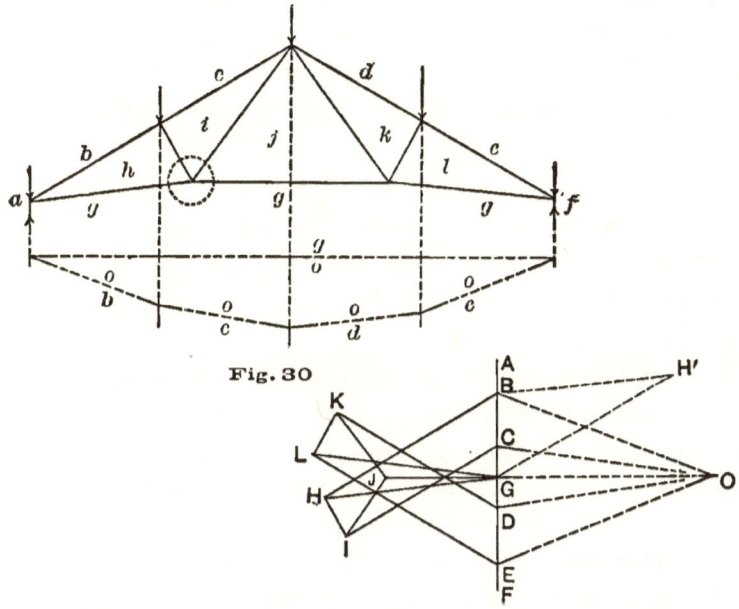

Fig. 30

We cannot, however, begin with the joint just considered, since at first the four forces are all unknown in magnitude.

If, however, we start with the joint *gabh*, the polygon of forces can be at once drawn. For, reasoning as above, it is seen that the portion of the truss immediately surrounding this joint is in equilibrium under the action of four forces: the reaction in the line *ga*, the load in the line *ab*, and the internal forces in the lines *bh*, *hg*. Of these forces, two (the reaction and the load) are completely known; and it is necessary only to draw a polygon of which two sides represent the known forces and the other two sides are made parallel to the members *bh*, *hg*. Such a polygon is shown in Fig. 30; and *BH* and *HG* represent in magnitude and direction the forces whose lines of action are *bh*, *hg*.

Evidently, BH and HG represent also the magnitudes (Art. 63) of the internal stresses in the two members bh and hg. The nature of these stresses may be found as follows: Since the four forces represented in the polygon $GABHG$ are in equilibrium, and since GA, AB are the directions of two of them, the directions of the other two must be BH, HG. Hence, BH acts *toward* the joint and HG *from* it. This shows that the stress in bh is a *compression*, while that in hg is a tension.

Passing now to the joint $bcihb$, it is seen that of the four forces whose lines of action meet there, two are fully known, namely, the load BC acting vertically downward and the internal force in hb acting toward the joint (since the stress is compressive), while the remaining two (viz., the internal forces in ci and ih) are unknown in magnitude and direction. Since, however, the unknown forces are but two in number, the force polygon can be completely drawn, and is represented by the quadrilateral $HBCIH$. The directions of the forces are found as in the preceding case, and it is seen that the bars ci and ih both sustain compressive stresses.

The process may be continued by passing to the remaining joints in succession, in such order that at each there remain to be determined not more than two forces. The complete construction is shown in Fig. 30.

It is evident that the loads AB and EF might have been omitted without changing the stresses in any of the truss members. For their omission would leave as the complete force polygon for external forces $BCDEGB$, and the two reactions would be GB and EG; but the force diagram would be otherwise unchanged.

The great convenience of the notation adopted is now seen.*

---

* This is known as Bow's notation. The notation adopted in Part I involves the same idea, but it is not usually employed in works on Graphic Statics, though possessing very evident advantages. It was suggested to the writer by its use in certain of Professor Eddy's works.

ROOF TRUSS WITH VERTICAL LOADS. 65

The line representing the stress in any member is designated in the force diagram by letters similar to those which designate that member in the truss diagram. The latter is evidently a *space diagram* (Art. 11). The force diagram is often called a *stress diagram*, since it shows the values of the internal stresses in the truss members.

89. **Reciprocal Figures.** — There are always two ways of completing the force polygon when two of the forces are known only in lines of action. (See Art. 18.) Either way will give correct results, but unless a certain way be chosen, it will become necessary to repeat certain lines in the stress diagram. Thus, if, in Fig. 30, instead of *GABHG* we draw *GABH'G*, the lines *GH'*, *H'B* are not in convenient positions for use in the other polygons of which they ought to form sides. The lettering of the diagrams will also be complicated. As an aid in drawing the lines in the most advantageous positions, it is convenient to remember the fundamental property of figures related in such a way as the force and space diagrams shown in Fig. 30.

Such figures are said to be *reciprocal* with regard to each other. The fundamental property of reciprocal figures is that for every set of lines intersecting in a point in either figure, there is in the other a set of lines respectively parallel to them and forming a closed polygon.

It is also an aid to remember that the order of the sides in any closed polygon in the stress diagram is the same as the order of the corresponding lines in the truss diagram, if taken consecutively around the joint. This usually enables us at once to draw the sides of each force polygon in the proper order.

90. **Order of External Forces in Force Polygon.** — It will be observed that in the case above considered, in constructing the force polygon for the loads and reactions, these forces have been taken consecutively in the order in which their points of

application occur in the perimeter of the truss. This is a necessary precaution in order that the stress diagram and truss diagram may be reciprocal figures, so that no line in the former need be duplicated.

This requirement should be especially noticed in such a case as that shown in Fig. 31, in which loads are applied at lower as

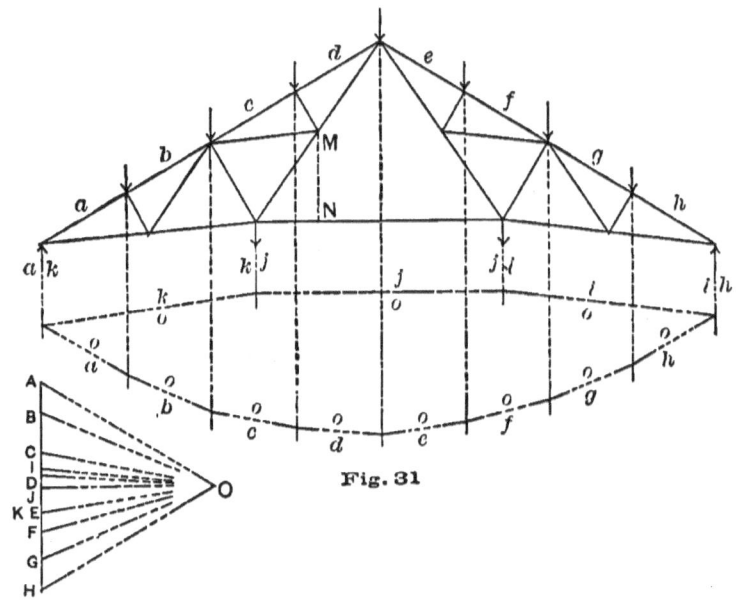

Fig. 31.

well as at upper joints. If the reactions are found by the method of Art. 87, without modification, the force polygon will not show the external forces in the proper order, since the known forces are not applied at consecutive joints of the truss. A new polygon should therefore be drawn after the reactions have been determined.

If desirable (as in some cases it may be) to make use of a funicular polygon in which the external forces are taken consecutively, this may be drawn after the reactions are determined and the new force polygon is drawn.

If a load be applied at some joint interior to the truss, as at M (Fig. 31), then in constructing the stress diagram it should be assumed to act at N, where its line of action intersects the exterior member of the truss, and the fictitious member MN inserted. The stresses in the actual truss members will be unaffected by this assumption, and such a device is necessary in order that the stress diagram may be the true reciprocal of the truss diagram.

§ 3. *Stresses Due to Wind Pressure.*

91. Direction of Reactions Due to Wind Pressure. — Since the effective pressure of the wind has the direction normal to the surface of the roof (Art. 85), it has a horizontal component which must be resisted by the reactions at the supports. These cannot, therefore, act vertically, as in the case when the loading is vertical. Their actual directions will depend upon the manner in which the ends of the truss are supported.

If the ends of the truss are immovable, the directions of the reactions cannot be determined, since any one of an infinite number of pairs of forces acting at the ends would produce equilibrium. (The same would be true of the reactions due to vertical loads.) In such a case the usual assumption is one of the following: (1) the reactions are assumed parallel to the loads; (2) the resolved parts of the reactions in the horizontal direction are assumed equal.

In the case of trusses of large span it is not unusual to support one end of the truss upon rollers so that it is free to move horizontally, the other end being hinged, or otherwise arranged to prevent both horizontal and vertical motion. This allows for expansion and contraction with change of temperature, as well as for movements due to the small distortions of the truss under loads. With this arrangement the reaction at the end supported on rollers must be vertical; and since the point of application of the other reaction is known, both can be fully determined by the method described in Art. 39.

92. Determination of Reactions. — The methods of finding reactions will now be explained for the three cases mentioned in the preceding article: (1) Assuming both reactions parallel to the wind pressure; (2) assuming the horizontal resolved parts of the two reactions equal; and (3) assuming one reaction vertical.

(1) The first case needs no explanation, since it is identical with that described in Art. 87, except that the loads and reactions have a direction normal to one surface of the roof, instead of being vertical. It is to be noticed that this assumption cannot be made if the roof surface is curved, since the lines of action of the forces will not be parallel. But since the direction of the resultant of the loads will be known from the force polygon, both reactions may be assumed to act parallel to this resultant, and the construction made as before.

(2) In the second case, let each reaction be replaced by two forces acting at the support, one horizontal and the other vertical. The two horizontal forces are known as soon as the force polygon for the loads is drawn, and the two vertical forces may be found as in the preceding case, since their lines of action are known.

In Fig. 32, let *ab*, *bc*, *cd* be the lines of action of the wind forces. Let the right reaction be considered as made up of a horizontal component acting in *de* and a vertical component acting in *ef*; and let the left reaction be replaced by a vertical component acting in *fg* and a horizontal component acting in *ga*. Draw the force polygon (or "load-line") *ABCD*. By the assumption already made *GA* and *DE* are to be equal, and their sum is to equal the horizontal resolved part of *AD*. Through the middle point of *AD* draw a vertical line; its intersections with horizontal lines through *A* and *D* determine the points *G* and *E*, so that the two forces *GA* and *DE* become known. The only remaining unknown forces are the vertical forces *EF* and *FG*. Choose a pole *O*, draw rays to the points *G*, *A*, *B*, *C*, *D*, *E*, and then the corresponding strings. Through the points determined by the intersection of *og* with *fg*, and *oe*

STRESSES DUE TO WIND PRESSURE. 69

with *cf*, draw the string *of*. The corresponding ray drawn from *O* intersects *EG* in the point *F*, thus determining *EF* and *FG*. The reactions are now wholly known; that at the left support being *FA*, and that at the right support *DF*.

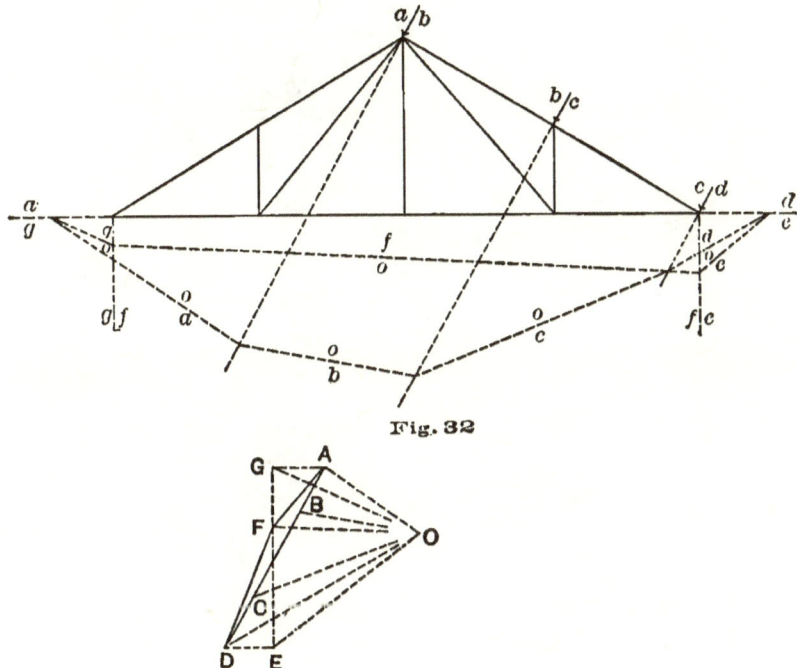

Fig. 32

(3) For the third case the construction is shown in Fig. 33 (*A*) and (*B*), for the two opposite directions of the wind. The method is identical with that employed in Art. 39. Only one point of the line of action of the left reaction is known, hence this is taken as the point of intersection of the corresponding strings of the funicular polygon. One of these strings can be drawn at once, since the corresponding ray is known; and the other is known after the remaining strings have been drawn, since it must close the polygon. The construction should be carefully followed through by the student.

The force polygons for the two directions of the wind are distinguished by the use of *O* and *O'* to designate the two poles.

70   GRAPHIC STATICS.

In Fig. 33 (A), ABCDEFGHIA is the force polygon for the case when the wind is from the right. Notice that the points A, B, C, D, coincide. This means that the loads AB, BC, CD, are each zero.

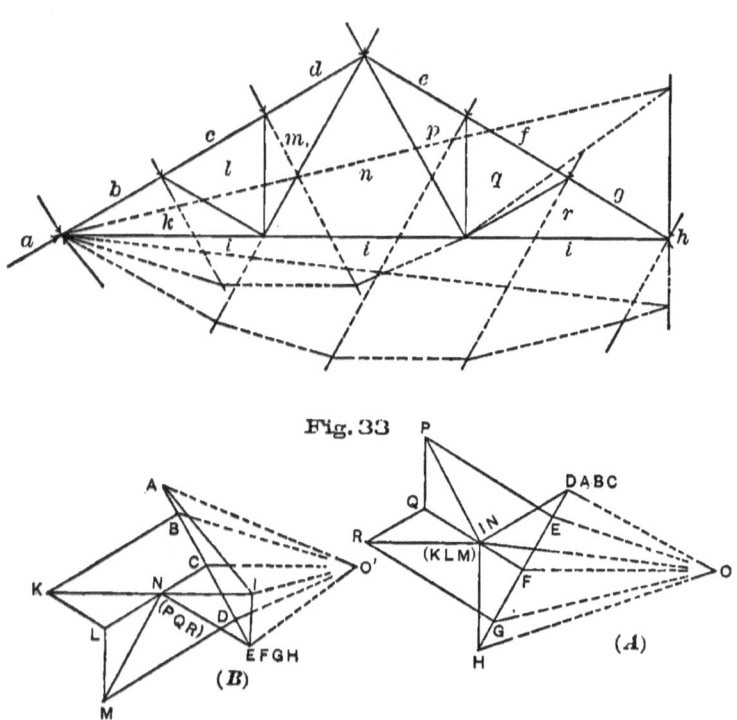

Fig. 33

In diagram (B), ABCDEFGHI A is the force polygon for the case when the wind is from the left. The points E, F, G, H, coincide, because the loads EF, FG, GH, are each zero.

93. **Stress Diagrams for Wind Pressure.** — When the loads and reactions due to wind pressure are known, the internal stresses can be found by drawing a stress diagram, just as in the case of vertical loads. The construction involves no new principle, and will be readily made by the student. In Figs. 33 (A) and 33 (B) are shown the diagrams for the two directions of the wind.

The stresses in all members of the truss must be determined for each direction of the wind. If the truss is symmetrical with respect to a vertical line, as is usually the case, it may be that the same stress diagram will apply for both directions of wind. This will be so if the reactions are assumed to act as in cases (1) and (2) of the preceding article. In the case represented in Fig. 33, however, the hinging of one end of the truss destroys the symmetry of the two stress diagrams, and both must be drawn in full.

§ 4. *Maximum Stresses.*

94. **General Principles.** — For the purpose of designing any truss member, it is necessary to know the greatest stresses to which it will be subjected under any possible combination of loads.

Stresses are combined in accordance with the principle stated in Art. 78, that the resultant stress in a truss member due to the combined action of any loads is equal to the algebraic sum of the stresses due to their separate action.

The method will be illustrated by the solution of an example with numerical data.

95. **Problem — Numerical Data.** — Let it be required to design a wrought iron truss of 40 ft. span, of the form shown in Fig. 34 (Pl. I). Let 12 ft. be the distance apart of trusses, and let the loads be as follows:

Weight of truss, to be assumed in accordance with the formula of Art. 82: $W = \frac{3}{4} al (1 + \frac{1}{10} l)$. This gives $W = 1800$ lbs. Assuming this to be divided equally among the upper panels, and that the load for each panel is borne equally by the two adjacent joints, the load at each of the joints *bc, cd, de* is 450 lbs. The loads at the end joints may be neglected, being borne directly at the supports.

Weight of roof. — This depends upon the materials used and the method of construction, but will be taken as 6 lbs. per sq. ft.

of roof area, giving 900 lbs. as the load at each joint. This, also, is a permanent load. Total permanent load per joint, 1350 lbs.

Weight of snow. — Taking this as 15 lbs. per horizontal square foot, we find 1800 lbs. as the load at each joint.

Wind pressure. — This is computed from the formula

$$p_a = \frac{2 \sin a}{1 + \sin^2 a} p_n. \qquad \text{(Art. 85.)}$$

For this case we put $\sin a = \frac{15}{25} = \frac{3}{5}$; $p_n = 40$ lbs. per sq. ft.; whence $p_a = 35$ lbs. per sq. ft. (about). This gives upon each panel of the roof 5250 lbs. Then with the wind from either side, the wind loads on that side would be 2625 lbs., 5250 lbs., 2625 lbs. respectively.

96. Stress Diagrams. — We are now ready to construct the stress diagrams.

The truss is shown (Pl. I) in Fig. 34 (*A*). Fig. 34 (*B*) is the stress diagram for permanent loads. No diagram for snow loads is needed, since it would be exactly similar to that for permanent loads. The snow load at any joint being four-thirds as great as the permanent load, the stress in any member due to snow is four-thirds that due to permanent loads.

Fig. 34 (*C*) shows the stress diagram for the case of wind blowing from the left. The reactions are assumed to act in lines parallel to the loads — that is, normal to the roof. With this assumption, no separate diagram is needed for the case of wind from the right, since such a diagram would be exactly symmetrical to Fig. 34 (*C*). For example, the stress in the member *gh* due to the wind blowing from the right is given by the line *GM* in Fig. 34 (*C*).

97. Combination of Stresses. — After the stress diagrams are completed for the various kinds of loads, the stresses should be scaled from the diagrams and entered with proper sign in a table, as follows:

MAXIMUM STRESSES.

Member.	Permanent Load.	Snow.	Wind R.	Wind L.	Max.
bh	− 4560	− 6080	− 6270	− 7925	− 18570
ci	− 3760	− 5010	− 6270	− 7925	− 16700
hi	− 1080	− 1440	0	− 5250	− 7770
ik	+ 1785	+ 2380	+ 550	+ 6650	+ 10820
gh	+ 3735	+ 4980	+ 2600	+ 8800	+ 17520
gk	+ 2190	+ 2920	+ 2300	+ 2300	+ 7410
gm	+ 3735	+ 4980	+ 8800	+ 2600	+ 17520
lk	+ 1785	+ 2380	+ 6650	+ 550	+ 10820
ml	− 1080	− 1440	− 5250	0	− 7770
dl	− 3760	− 5010	− 7925	− 6270	− 16700
em	− 4560	− 6080	− 7925	− 6270	− 18570

By combining the results, the maximum stress in each member for any possible condition of loading can be determined. The possible combinations of loading are the following: Permanent load alone; permanent and snow loads; permanent load, and wind from either direction; permanent and snow loads, and wind from either direction. The student will readily understand the method of combining the separate results.

The problem here solved relates to a very simple form of truss. With some forms there may occur a reversal of stress in certain members, under different conditions of loading.

It is to be noticed that in the table the word maximum is used in its numerical sense, and has no reference to the algebraic sign of the stress.

98. Examples. — In Fig. 35 (*A*) to (*F*), are shown several forms of truss for which the student may draw stress diagrams, assuming loads in accordance with the data given in Arts. 82 to 85. In determining reactions due to wind pressure, the

74 GRAPHIC STATICS.

three assumptions mentioned in Art. 92 should all be used in different cases, that the student may become familiar with the principle of each.

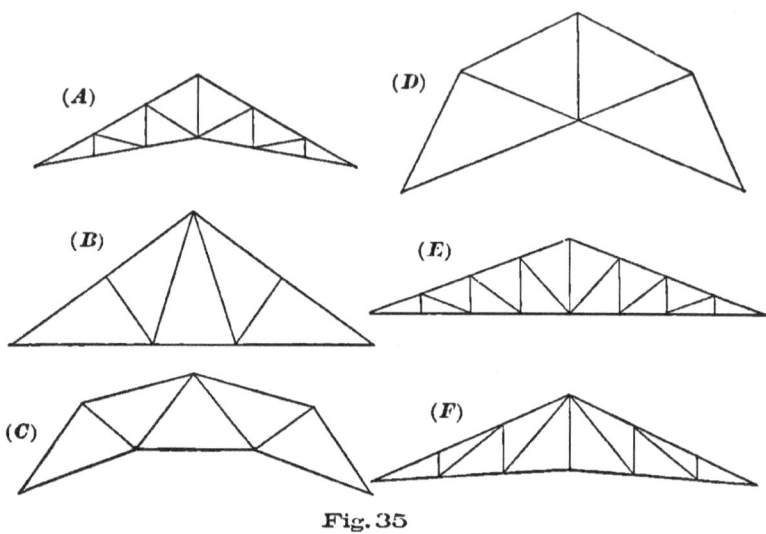

Fig. 35

§ 5. *Cases Apparently Indeterminate.*

99. Failure of Usual Method. — In attempting to construct the stress diagram by drawing the force polygon for each joint in succession, as in the cases thus far treated, a difficulty is met in certain forms of truss. It may happen that after proceeding to a certain point it is impossible to select a joint for which the force polygon can be completely drawn, the number of unknown forces for every joint being greater than two.

Thus, in the truss shown in Fig. 36, if the stress diagram is started in the usual way, beginning at the left support, the force polygons for three joints may be constructed without difficulty, thus determining the stresses in *bl*, *lk*, *lm*, *cm*, *mn*, *nk*. But the force polygon for *cdqpnmc* cannot be constructed, since three forces are unknown, — namely, those in *dq*, *qp*, *pn*. And at the joint *knpsk*, the stresses in *np*, *ps*, and *sk* are unknown. The problem, therefore, seems at this point to become indeter-

CASES APPARENTLY INDETERMINATE.

minate, since either of the two polygons can be completed in any number of ways, so far as the known forces determine. It can be shown, however, that this ambiguity is only apparent. This may be proved as follows :

Consider the portion of the truss to the left of the broken line $M'N'$. It is in equilibrium under the action of eight forces ; five of these (four loads and a reaction) are known ; the remaining three are the forces in cr, rs, and sk. Now, the problem of determining these three unknown forces is the same as that treated in Art. 40. It was there found to be a determinate problem, unless the lines of action of the three unknown forces intersect in a point or are parallel.

That the problem is determinate may be seen also from the principle of moments (Art. 51). The eight forces mentioned being in equilibrium, the sum of their moments is zero for any origin in their plane. Let the origin be taken at the point of intersection of the lines of action of two of the unknown forces, as cr, rs. Then from the principle of moments we have (since the moments of the two forces named are zero) : Algebraic sum of moments of loads and reaction to left of section + moment of $SK = 0$. The only unknown quantity in this equation is the magnitude of SK, which may, therefore, be determined. The other unknown forces may be found in a similar manner, the origin of moments being in each case chosen so as to eliminate two of the three unknown forces.

The whole problem of drawing the stress diagram is now seen to be determinate. For, as soon as the stress in sk is known, the force polygon for the joint $knpsk$ contains but two unknown sides, and can be drawn at once. No further difficulty will be met.

100. Solution of Case of Failure — First Method. — The reasoning of the preceding article suggests two methods of treating the so-called ambiguous case. These will now be described.

76   GRAPHIC STATICS.

The first method is to apply the construction of Art. 40, as follows: Referring to Fig. 36, consider the equilibrium of the portion of the truss to the left of the line $M'N'$. The system of forces consists of those whose lines of action are *ka, ab, bc, cd, de, er, rs, sk*. Let them be taken in the order named, and draw the force polygon for the known forces. (The reaction *ka* is supposed to be already determined.) The known part of the force polygon is $KABCDE$; the unknown part is to be marked $ERSK$. Choose a pole O and draw rays to K, A, B, C, D, and E; then draw the corresponding strings of the funicular polygon. Remembering the method of Art. 40, we draw first *oe*, making it pass through the intersection of *er* and *rs*

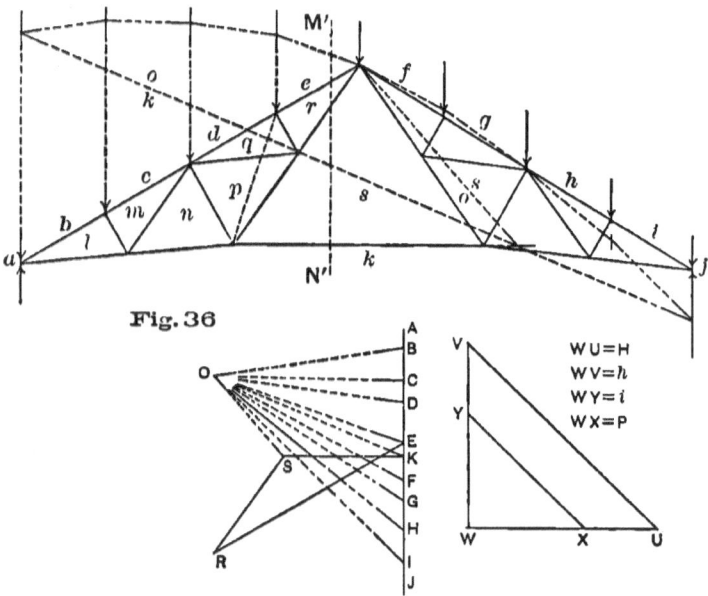

Fig. 36

(the reason for this being that *it makes the string os also pass through that point*); then draw in succession *od, oc, ob, oa, ok*. The string *ok* intersects *sk* in a point through which *os* must be drawn. Hence *os* must join that point with the starting point of the polygon. This completes the funicular polygon, except the string *or*, which must pass through the intersection of *er* and *rs* in some direction not yet known. This string is not necessary to the solution of the problem.

Since *os* is now known, *OS* may be drawn parallel to it from *O*; and by drawing a line from *K* parallel to the known direction of *KS*, the point *S* is determined, and *KS* becomes known.

To find *ER* and *RS* it is only necessary to draw from *S* a line parallel to *rs*, and from *E* a line parallel to *er*; their intersection gives the point *R*, and the force polygon for the system of forces considered is complete. The stresses in *er*, *rs*, and *sk* being now known, the stress diagram may be completely drawn by the usual method.

It is interesting to notice that the method just described determines the lines *ER*, *RS*, *SK*, in their proper position in the complete stress diagram. The determination of *ER* and *RS* by this method is not necessary, since the usual method of drawing the stress-diagram can be carried out as soon as *SK* is known.

The funicular polygon employed in the above construction, so far as it belongs to the external forces, may coincide with the corresponding part of the funicular polygon used in determining the reactions. If this is desired, two points must be observed: (1) the string *oe* must be the first drawn, and must pass through the intersection of *er* and *rs*, and (2) the pole must be so chosen that *ok* will not be nearly parallel to *ks*.

It should also be noticed that the construction of the funicular polygon might begin with the string *ok*, which should then be made to pass through the intersection of *rs* and *sk*. The student will be able to carry out this construction without difficulty.

101. **Solution of Case of Failure — Second Method.** — It will now be shown how the apparently ambiguous case can be

treated graphically by the principle of moments. Referring again to Fig. 36, consider the portion of the truss to the left of section $M'N'$. It is acted upon by eight forces, of which five (the loads and the reaction) are known, and three (whose lines of action are *er*, *rs*, *sk*) are unknown. The algebraic sum of the moments of all these forces about any origin must be zero. Let the origin be taken at the point of intersection of *er* and *rs*, so that the moments of the forces acting in these lines are both zero; then the sum of the moments of the five known forces, plus the moment of the force acting in the line *sk*, must equal zero. Now the sum of the moments of the five known forces may be found by the method of Art. 56. Through the origin of moments draw a line parallel to the resultant of the forces named (that is, a vertical line), and let i equal the length intercepted on it by the strings *oe*, *ok*. Then the required moment is $-iH$, where H is the pole distance. (The minus sign is given in accordance with the convention that left-handed rotation shall be positive.) Let $P=$ unknown force in line *sk*, and h its moment-arm. For the purpose of computing the moment, assume the stress in *sk* to be a tension; then the force P acts toward the right and its moment is positive, the value being $+Ph$.

Hence, $\qquad Ph - Hi = 0$;

or, $\qquad\qquad P = \dfrac{i}{h} H.$

From this equation P may be computed. The computation may be made graphically as follows: Draw (Fig. 36) a triangle WUV, making $WU = H$ (force units) and $WV = h$ (linear units). Lay off $WY = i$ (linear units) and draw YX parallel to VU. Then WX (force units) represents P. This is readily seen, since from the two similar triangles we have the proportion $\dfrac{P}{H} = \dfrac{i}{h}$, which agrees with the equation above deduced.

The computation is simplified if the pole distance H is taken equal to as many force units as h is linear units; or if H

CASES APPARENTLY INDETERMINATE. 79

is some simple multiple of h. For, suppose $H = nh$; then $P = \dfrac{inh}{h} = ni$. If $n = 1$, $P = i$.

The stress in sk is found to be a tension, since P is positive. Whatever the nature of the stress, it may be assumed a tension in writing the equation, and the sign of the value found will show whether the assumption coincides with the fact.

102. **Other Methods for Case of Failure.** — In certain cases the method of treating the "ambiguous case" may profitably be varied.

(1) The construction of Art. 100 may be modified as follows: Determine the resultant of the known external forces acting on the portion of the truss to one side of the section $M'N'$. This resultant is in equilibrium with the three unknown forces acting in the members er, rs, sk. Hence these forces can be determined by the special method explained in Art. 42.

The resultant of the five known forces is represented in magnitude and direction by KE in the force polygon; and its line of action passes through the intersection of the strings oe, ok in the funicular polygon. Since this point of intersection is likely to be inaccessible, the construction of Art. 42 cannot be conveniently applied. It may be modified by using instead of KE the two forces KA (the reaction in the line ka) and AE (the resultant of the four loads, its line of action being determined by the intersection of the strings oa and oe). First determine forces in the three lines er, rs, sk, which shall be in equilibrium with KA; then make a similar construction for AE, and combine the results.

(2) It has been proposed to employ the following reasoning: Remove the members pq and qr, and insert another represented by the broken line in Fig. 36. Evidently this does not change the stress in the member sk, since the forces acting on the truss to the left of the section $M'N'$ are unchanged. But with this change the difficulty encountered in constructing the stress diagram by the usual method is avoided. For when the joint

nmcdqpn is reached, the forces acting there will be all known except two. Let the stress diagram be drawn in the usual way until the stress in *sk* is known. Then restore the original bracing and repeat the construction, using the value just determined for *SK*.

This method is convenient whenever it is applicable. Cases may, however, arise, in which it will fail. For instance, if a load is applied at the joint *pqrsp*, the members *pq* and *qr* cannot both be omitted, and the method cannot be applied. In such cases, one of the methods explained in the preceding articles may be applied.

Other methods might be mentioned, but the foregoing discussion of the case will probably be found sufficient.

103. **Failing Case in Other Forms of Truss.** — The usual method of constructing the stress diagram may fail in other forms of truss, though the one above described is the most common. In such a case, the problem of finding the stresses may be really indeterminate, or only apparently so. Whenever it is possible to divide the truss into two parts by cutting three members which are not parallel and do not intersect in one point, the stresses in the three members cut are determinate as soon as all external forces are known, and can be found by methods already given. If more than three members are cut, the problem of finding the stresses in them is indeterminate, unless all but three of these stresses are known. By remembering these principles, the determinateness of any given problem may readily be tested. (See Art. 70.)

§ 6. *Three-Hinged Arch.*

104. **Arched Truss Defined.** — If a truss is so supported that when sustaining vertical loads the reactions at the supports have horizontal components directed toward the center of the

span, the truss becomes an *arch*. The only kind of arch we shall here consider is that consisting of two partial trusses hinged together at the crown, and each hinged at the point of support.

Such a truss is shown in Fig. 38, in which the two partial trusses are hinged to the abutments at P and Q, and connected by a hinge at the point R. Since a hinge at the support allows the reaction of the supporting body upon the truss to take any direction in the plane of the truss, the directions of the reactions at P and Q are unknown, as is also that of the force exerted by either partial truss upon the other at the point R. The problem of determining the reactions may, therefore, at first sight seem indeterminate. It will be shown in the next article that it is in reality determinate, and that the only principles needed in the solution are such as have been already often applied in the preceding chapters. The three-hinged arch is indeed a "simple" structure (Art. 75), since the theory of elasticity is not needed in the determination of the reactions.

105. Reactions Due to a Single Load.—The method of finding the reactions is most clearly understood by considering the

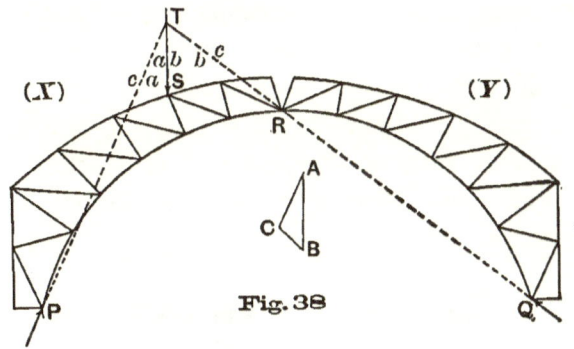

Fig. 38.

effect of a single load on either partial truss. Let a load be applied at S (Fig. 38) and let all other loads acting on either portion of the truss (including the weight of the structure) be neglected. Call the two partial trusses X and Y, and consider the part Y. The only forces acting upon it are the reaction exerted by the abutment at Q and the force exerted at R by the truss X. These two forces, being in equilibrium, must have the same line of action, which is, therefore, the line QR. Consider now the body X. The external forces acting upon it are the load at S, the reaction of the abutment at P, and the force exerted by Y at the point R. But this last force is equal and opposite to the force exerted by X upon Y, and its line of action is therefore QR. The three forces acting upon the body X being in equilibrium, their lines of action must meet in a point; which point is found by prolonging QR to meet the line of action of the applied load. Let T be this point, then PT is the line of action of the reaction at P. The reactions can now be determined by drawing the triangle of forces. This triangle is ABC in the figure, AB representing the load at S, BC the reaction in the line QR (also marked bc), and CA the reaction in the line PT (marked also ca). Evidently ABC may be regarded as the polygon of external forces, either for the partial truss X, or for the whole structure composed of X and Y; and BC represents either the force exerted by Y upon X at R, or the force exerted upon Y by the supporting body at Q.

If, now, the structure be loaded at other points, the reactions due to each load may be found separately; the resultant of all such separate reactions at either support will be the true reaction at that support when all the loads act together. A convenient method of applying these principles will be given in the next article.

106. **Reactions and Stresses Due to Any Vertical Loads.** — In Fig. 39 is represented a truss consisting of two parts supported by hinges at P' and Q' and hinged together at R'. Consider all vertical loads to be applied at the upper joints, their lines of action being marked in the usual way. We shall first explain the construction for finding the reactions at the supports; after

these are determined, the drawing of the stress diagram will present no difficulty.

Since we shall sometimes deal with one of the partial trusses, and sometimes with the two considered as a single body, it will be well at the outset to specify the external forces acting on each of these bodies.

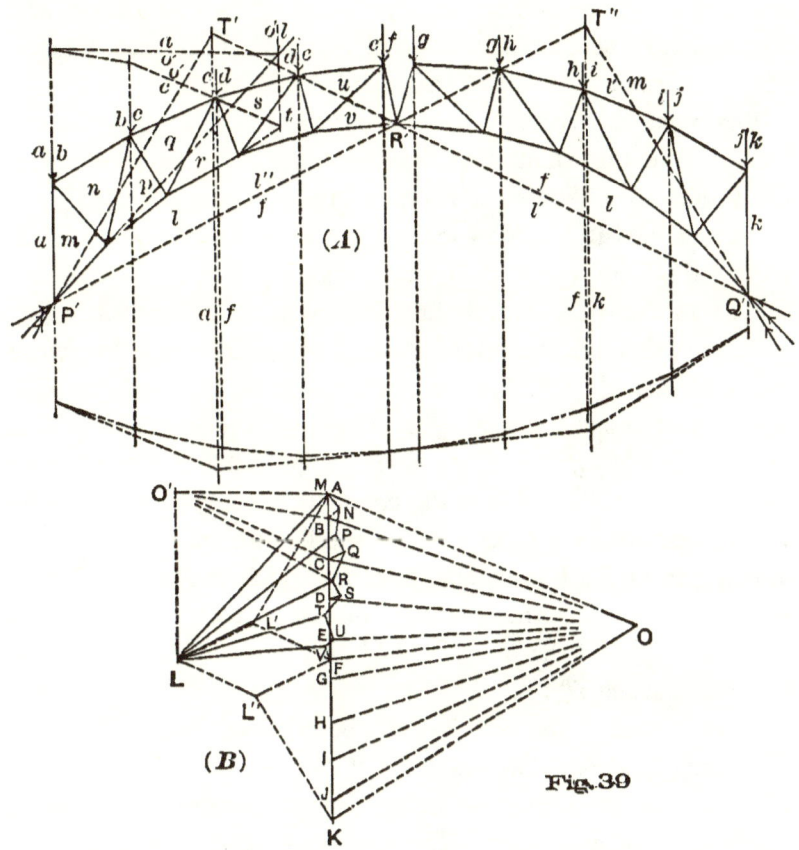

Fig. 39

(1) For the partial truss at the left we have five applied loads, the reaction at R' (exerted by the other truss), and the reaction at P'. The force polygon for these forces will be marked as follows: *ABCDEFLA*. (The meaning of the letters will be understood before the force polygon is actually drawn, by reference to the corresponding letters in the space diagram.)

(2) For the right-hand partial truss the external forces are five loads and two reactions, and the force polygon will be marked thus: *FGHIJKLF*. (Notice that *FL* and *LF* are equal and opposite forces, being the "action and reaction" between the two trusses at R'.)

(3) For the combined structure the external forces are the ten loads and the reactions at P' and Q'. (The action and reaction at R' become now internal forces.) The force polygon will be marked thus: *ABCDEFGHIJKLA*.

Begin the construction by drawing the force polygon for the ten loads on the whole structure, lettering it as just indicated. Choose a pole O, draw the rays, and then the funicular polygon as far as possible. Now consider the right partial truss as unloaded. The resultant of the remaining five loads is represented in magnitude and direction by *AF*; its line of action must pass through the intersection of *oa* and *of*, and is therefore the line marked *af*. Now, reasoning as in the preceding article, we see that the reaction at Q' must act in the line $Q'R'$. Let $Q'R'$ intersect *af* in T', then $P'T'$ is the line of action of the reaction at P'. Hence the complete force polygon for the whole truss, when the right half is unloaded, may be found by drawing from F a line parallel to $Q'R'$, and from A a line parallel to $P'T'$, prolonging them till they intersect at L'. The reaction at P' is $L'A$, and that at Q' is FL'. (The line of action of the latter is marked *fl'*.)

Next, consider the left half to be without loads, the other half being loaded. The resultant of the five loads now acting is *FK*, its line of action *fk* being drawn through the point of intersection of *of* and *ok*. The reactions at P' and Q' for the present case have lines of action $P'R'$ and $Q'T''$, found just as in the first case of loading. These reactions are therefore determined in magnitude and direction by drawing from K a line parallel to $Q'T''$ and from F a line parallel to $P'R'$, prolonging them till they intersect at L''. The complete force polygon for this case of loading is therefore *FGHIJKL''F*.

Consider now that both parts of the truss are loaded. The reaction at P' is the resultant of the two partial reactions $L'A$, $L''F$, and the reaction at Q' is the resultant of the two partial reactions FL', KL''. From L'' and L' draw lines parallel respectively to FL' and FL'', intersecting in L. Then $L''L$ is equal and parallel to FL', and LL' is equal and parallel to $L''F$. Hence KL and LA represent the resultant reactions at Q' and P' respectively. This completes the polygon of external forces for the whole truss, as well as that for each partial truss.

The stress diagram can now be drawn in the usual way, beginning at the point P'. The diagram for one partial truss is shown in Fig. 39 (*B*).

If loads are applied at the lower joints of the trusses, the reactions due to these may be found in the same manner as for the upper loads. But before beginning the determination of the stresses, the polygon must be drawn for the external forces taken in order around the truss. (See Art. 90.)

107. Case of Symmetrical Loading. — If the two half trusses are exactly similar and symmetrically loaded, the determination of reactions and stresses is much simplified.

(1) As regards the reactions, symmetry shows that the forces exerted by the two trusses upon each other at R' are horizontal. Hence, referring to Fig. 39, and considering either half-truss, as that to the left, the line of action of the reaction at P' may be found by drawing a horizontal line through R' and prolonging it to intersect *af*, the line of action of the resultant of all loads on the left truss; the line joining this point of intersection with P' is the required line of action of the reaction at the left abutment. The two reactions are now determined by drawing from F a horizontal line and from A a line parallel to the line just determined, and prolonging them till they intersect.

(2) As to the stresses, only one partial truss need be considered, since the stress diagrams for the two portions will be symmetrical figures.

86 GRAPHIC STATICS.

These principles might have been employed in the case discussed in the preceding article; but the method there used is applicable to cases in which either the trusses or the loads are unsymmetrical.

108. **Wind Pressure Diagram.** — The diagrams for wind pressure will present no difficulty. The determination of the reactions will, indeed, be simpler than in the preceding case, since only one partial truss will be loaded at any one time, and the line of action of one reaction is therefore known at the outset. The construction is shown in Fig. 40.

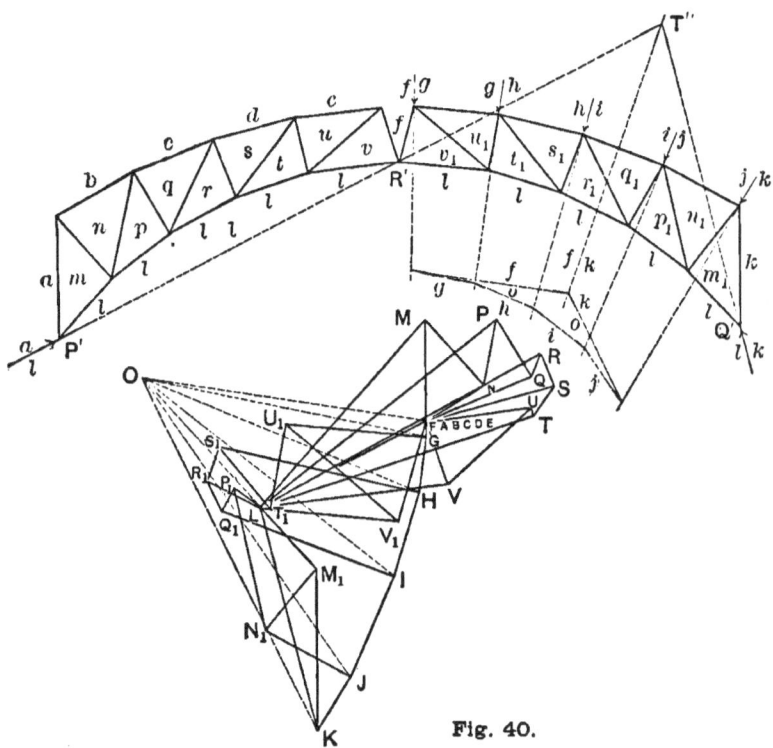

Fig. 40.

In computing wind pressure loads, it will be assumed that each joint sustains half the pressure coming upon each of the two adjacent panels. It will be sufficiently correct in computing the pressure at any joint, to assume the slope of the roof as that of the tangent to the roof curve at the joint in question; and the direction of the wind load may be taken as that of the normal to the roof curve at the joint. The force polygon for the wind loads is therefore not a straight line when the roof is curved. In Fig. 40, this polygon is $FGHIJK$. The reactions are to be marked KL, LF. (Since there are no loads on the other half-truss, the points $ABCDEF$ will coincide.) Choosing a pole O, draw the funicular polygon as shown, and determine the line of action of the resultant of all the wind loads. This line of action fk is drawn parallel to FK, through the point of intersection of the strings of and ok. Prolong fk to intersect $P'R'$ produced at T''; then $Q'T''$ is the line of action of the reaction at Q'. The force polygon may now be completed by drawing KL parallel to $Q'T''$ and LF parallel to $P'R'$. The reactions KL and LF being thus determined, the stress diagram can be drawn without difficulty.

The stress diagram is drawn for both partial trusses, with the result shown in the figure. If the two trusses are symmetrical, the diagram for the other direction of the wind need not be drawn; the stresses for this case being found from the diagram already drawn. If, however, the partial trusses are dissimilar, a second wind-diagram must be drawn.

It is not necessary to draw separate space diagrams for vertical loads and wind-forces. The constructions of Figs. 39 and 40 have been here kept wholly separate, in order that the explanation may be more easily followed.

109. **Check by Method of Sections.** — In case of a truss of long span, especially when the members have many different directions and are short compared with the whole length of the span, the small errors made in drawing the stress diagram are likely to accumulate so much as to vitiate the results. Thus, in Fig. 39, if, in drawing the stress diagram, we begin at P' and pass from joint to joint, there is no check upon the correctness of the work until the point R' is reached. At that point we

have a second determination of the reaction exerted by each half-truss upon the other; and it is quite likely that the two values found will not agree.

By a method like that employed in Art. 100 for the "indeterminate" form of truss, we may avoid the necessity of making so long a construction before checking the results. In Fig. 39 take a section cutting the three members cq, qr, rl, and apply the principles of equilibrium to the body at the left of the section. The forces acting on this body are LA, AB, BC, CQ, QR, RL. Choose a pole O' and draw the funicular polygon for this system, making the string $o'c$ pass through the point of intersection of cq and qr. Draw successively the strings $o'c$, $o'b$, $o'a$, $o'l$, prolonging the last to intersect lr. Through the point thus determined, draw $o'r$, which must also pass through the point of intersection of cq and qr. As soon as $o'r$ is known, the corresponding ray $O'R$ can be drawn in the force diagram, and then by drawing from L a line parallel to lr, the point R is determined. We may now close the force polygon, since the directions of the two remaining forces (cq and qr) are known.

The stresses in the three members cut being now known, we have a check on the correctness of the construction of the stress diagram, as soon as these members are reached in the process.

This method will be found of great use, not only for this form of truss, but for any truss containing many members.

§ 7. *Counterbracing.*

110. **Reversal of Stress.** — If the loads supported by a truss are fixed in position and unchanging in amount, the stress in any member remains constant in magnitude and kind. But in most cases such are not the conditions, and it may happen that under different combinations of loading, the stress in a certain member is sometimes tension and sometimes compression. It

is often thought desirable to prevent such changes of stress, the design of the members and their connections being thereby simplified. To accomplish this is the object of *counterbracing*.

111. **Counterbracing.** — Consider a truss such as the one shown in Fig. 41, subjected to vertical loads and to wind pressure from either side. A diagonal member such as xy may sustain tension under vertical loads alone, or with the wind from the left; while with the wind from the right it may sustain compression.

Now suppose xy removed, and a member represented by the broken line xy' introduced. It may easily be shown that any

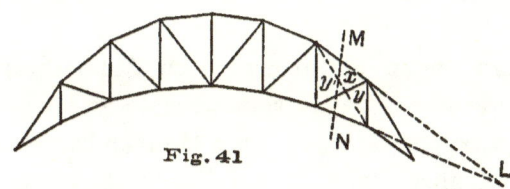

Fig. 41

system of loading which would cause compression in xy will cause tension in this new member; and *vice versa*. For, divide the truss by a section MN cutting xy and two chord members as shown, and let L be the point of intersection of the two chord members produced. The kind of stress in xy or the other member (whichever is assumed to be present) may be determined by considering the system of forces acting on one portion of the truss, as that to the left of the section MN. Let the principle of moments be applied to this system, the origin being taken at L. If the external forces acting on the portion of the truss considered be such as to tend to cause right-handed rotation about L, the stress in xy must be compression in order to resist this tendency; while, if xy be replaced by xy', a tension must exist in that member to resist right-handed rotation about L. Similarly, a tendency of the external forces to produce left-handed rotation about L would be resisted by a tension in xy; or by a compression in the member xy'.

If the two members act at the same time, the stresses in

them will be indeterminate. These stresses may, however, be made determinate by the following device:

Let the member xy be so constructed that it cannot sustain compression. Then, whenever the external forces are such as to tend to throw compression upon it, it ceases to act as a truss-member, and the member xy' receives a tension which is determinate.

If the member xy' be constructed in the same way, any tendency to throw compression upon it causes it immediately to cease to act, and puts upon the member xy a tension instead.

A member such as xy', constructed in the manner mentioned, is called a *counterbrace*.

112. Determination of Stresses with Counterbracing.

— The use of counterbracing adds somewhat to the labor of determining the maximum stresses, since the members actually under stress are not always the same. The method of treating such cases will be illustrated in the next article by the solution of an example; but first the main steps in the process may be outlined as follows:

(*a*) Construct separate stress diagrams for vertical loads and for wind in each direction, assuming the diagonals in all panels to slope the same way.

(*b*) Determine by comparison of these diagrams in which of the diagonal members the resultant stress is ever liable to be a compression. Draw in counters to all such members.

(*c*) With these counterbraces substituted for the original members, either draw new stress diagrams, or make the necessary additions to those already drawn. If the latter method is adopted, the added lines should be inked in a different color from that employed originally. (In some cases, this construction may be unnecessary on account of symmetry, as will be illustrated in the next article.)

(*d*) Combine the separate stresses for maxima in the usual way.

113. **Example.** — In Fig. 42 (Pl. II) are given stress diagrams for a "bow-string" roof-truss in which counterbracing is employed. At (A) is shown the truss or space diagram. The span is 48 ft.; rise of top chord, 16 ft.; rise of bottom chord, 8 ft. The chords are arcs of parabolas. The whole truss is divided into six panels by equidistant vertical members. The distance apart of trusses is taken as 12 ft.

Assume the weight of the truss at 2400 lbs., and that of the roof at 3600 lbs.; this makes the total permanent load 1000 lbs. per panel. Take 800 lbs. as the load at each upper joint, and 200 lbs. as the load at each lower joint.

The snow load, computed at 15 lbs. per horizontal square foot, is about 1440 lbs. per panel.

Wind pressure is to be computed from the formula of Art. 85.

We now proceed to apply the method outlined in the preceding article.

(a) Assuming one set of diagonals present, we construct the stress diagrams for the various kinds of loads.

Diagram for permanent loads. — This is shown at (B) Fig. 42, which needs little explanation. It will be noticed that the stress in every diagonal member is zero. This will always be the case if the chords are parabolic and the vertical loads are equal and spaced at equal horizontal distances.

Diagram for snow loads. — Fig. 42 (C) is the stress diagram for snow loads. In this case, also, the stresses in the diagonals are all zero.

Wind diagrams. — At (D) and (E) are shown the diagrams for the two directions of the wind. The only thing needing special mention is the method used in laying off the force-polygon for the wind loads. We first compute the normal wind pressure on each panel by the formula of Art. 85. We thus find, when the wind is from the right, the following total pressures, taking $p_n = 40$ lbs. per sq. ft.:

On panel *d*, 1650 lbs. On panel *e*, 3870 lbs. On panel *f*, 5480 lbs.

GRAPHIC STATICS.

In Fig. 42 (*D*) these are laid off successively in their proper directions to the assumed scale. Thus CD', $D'E'$, $E'G$ represent respectively 1650 lbs. normal to dq, 3870 lbs. normal to es, and 5480 lbs. normal to ft. Now each of these loads is to be equally divided between the two adjacent joints. Bisect CD' at D, $D'E'$ at E, and $E'G$ at F; then CD, DE, EF, FG represent the loads at the joints cd, de, ef, and fg respectively. The "load line" is therefore $CDEFG$.

The reactions are assumed to be parallel to the resultant load. With this explanation the figures (*D*) and (*E*) will be readily understood.

(*b*) *Comparison of results.* — The stresses due to permanent loads, wind right, and wind left, are shown in tabular form for convenience of comparison.

Member.	Perm. Load.	Snow Load.	Wind R.	Wind L.	Max.
aj	− 6660	− 9680	− 4250	− 14070	− 16340
bl	− 5350	− 7780	− 4880	− 9650	− 13130
cn	− 4550	− 6640	− 6380	− 6200	− 11190
dq	− 4550	− 6640	− 9500	− 4250	− 14050*
es	− 5350	− 7780	− 15030	− 3450	− 20380*
ft	− 6660	− 9680	− 14070	− 4300	− 20730*
jh_6	+ 5080	+ 7400	+ 780	+ 13500	+ 12480
kh_5	+ 4680	+ 6820	+ 700	+ 12450	+ 11500
mh_4	+ 4470	+ 6520	+ 1950	+ 7350	+ 10990
ph_3	+ 4470	+ 6520	+ 4100	+ 4100	+ 10990*
rh_2	+ 4680	+ 6820	+ 7680	+ 2050	+ 12360*
th_1	+ 5080	+ 7400	+ 13500	+ 800	+ 18580*
jk	+ 1180	+ 1440	+ 150	+ 2650	+ 2620
lm	+ 1180	+ 1440	− 280	+ 4100	+ 2620
np	+ 1180	+ 1440	− 950	+ 3800	+ 2620*
qr	+ 1180	+ 1440	− 1650	+ 2525	{ + 2620* / − 470* }
st	+ 1180	+ 1440	− 1350	+ 1250	{ + 2620* / − 170* }
kl	0	0	+ 1300	− 4600	+ 1300*
mn	0	0	+ 2700	− 4100	+ 2700*
pq	0	0	+ 4850	− 3150	+ 4850*
rs	0	0	+ 7080	− 1950	+ 7080*

It is seen that the diagonals shown are all in tension when the wind is from the right, and all in compression when the wind is from the left. Therefore counters are needed in all panels, and the counters will all come in action whenever the wind is from the left.

(*c*) *Stresses in counterbraces.* — It is evident from symmetry that no new diagrams are needed to determine the stresses in the counterbraces. In fact, the counterbrace in each panel is situated symmetrically to the main diagonal in another panel, and is subject to exactly equal stresses.

(*d*) *Combination for maxima.* — In combining the results for the greatest stresses in the various members, it will be assumed that the greatest snow load and the greatest wind load can never act simultaneously. For each member, therefore, the stress due to permanent load is to be added to the snow stress or the wind stress, whichever is greater. Again, in combining the tabulated results, we consider only the columns headed permanent load, snow load, and wind right; since whenever the wind is from the left, the other system of diagonals is in action. This gives the results entered in the last column.

We now notice the following facts:

(1) The results given for the diagonal members are the true maximum stresses.

(2) The stress found for each diagonal applies also to the symmetrically situated counterbrace.

(3) For any other member, we are to choose between the maximum found for that member and the value found for the symmetrically situated member. Thus, -20730 is the true maximum stress for both *aj* and *ft*; etc. (The numbers denoting true maximum stresses are marked with a (*) in the table.)

It is seen that the verticals, with one exception, may sustain a reversal of stress. Thus, *jk* and *st* must be designed for a tension of 2620 lbs., and also for a compression of 170 lbs.; while *lm* and *qr* are each liable to 2620 lbs. tension and to 470 lbs. compression.

CHAPTER VI. SIMPLE BEAMS.

§ 1. *General Principles.*

114. Classification of Beams. — A beam has been defined in Art. 79. Beams may be treated in two main classes, the basis of classification being that described in Art. 75. These two classes will be called *simple* and *non-simple* beams respectively. The present chapter deals only with simple beams, the definition of which may be stated as follows:

A *simple beam* is one so supported that it may be regarded as a rigid body in determining the reactions.

A simple beam may rest on two supports at the ends; or it may overhang one or both supports.

A *cantilever* is any beam projecting beyond its supports. Such a beam may be either simple or non-simple.

A *continuous beam* is one resting on more than two supports. Such a beam is *non-simple*.

A beam may be supported in several ways. It is *simply supported* at a point when it rests against the support so that the reaction has a fixed direction. It is *constrained* at a point if so held that the tangent to the axis of the beam at that point must maintain a fixed direction. If *hinged* at a support, the reaction may have any direction. We shall deal mainly with the case of simple support.

In what follows it will be assumed that the beam rests in a horizontal position, since this is the usual case.

115. External Shear, Resisting Shear, and Shearing Stress. — The *external shear* at any section of a beam is the algebraic

GENERAL PRINCIPLES.

sum of the external vertical forces acting on the portion of the beam to the left of the section.

The *resisting shear* at any section is the algebraic sum of the internal vertical forces in the section acting on the portion of the beam to the left, and exerted by the portion to the right of the section.

The *shearing stress* at any section is the stress which consists of the internal vertical forces in the section, exerted by the two portions of the beam upon each other. It consists of the resisting shear and the *reaction* to it. (See Art. 63.)

Let AB (Fig. 43) be a beam in equilibrium under the action of any external forces. At any point in its length, as C, conceive a plane to be passed perpendicular to the axis of the beam, and consider the portion AC, to the left of the section.

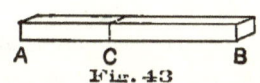

Fig. 43

The principles of equilibrium apply to the body AC, and the external forces acting upon it include, besides those forces to the left of C that are external to the whole bar, certain forces acting across the section at C that are *internal* to AB, but *external* to AC. (Art. 61.) These latter forces comprise that constituent of the internal stress between AC and CB which is exerted *by CB upon AC*.

Represent by V the algebraic sum of the resolved parts in the vertical direction of all forces acting on AB to the left of the section at C, upward forces being called positive. V is the *external shear* at the given section as above defined.

Since the body AC is in equilibrium, condition (1), Art. 58, requires that the algebraic sum of the resolved parts in the vertical direction of all forces acting on it must equal zero. Hence the forces acting on AC in the section at C must have a vertical component equal to $-V$. This vertical component is called the *resisting shear* in the given section. This resisting shear is one of the forces of a stress of which the other is an equal and opposite force exerted by AC upon CB. This stress is called the *shearing stress* in the section, and is called positive when it resists a tendency of AC to move upward, and of CB to move downward.

116. Bending Moment, Resisting Moment, and Stress Moment. — The *bending moment* at any section of a beam is the algebraic sum of the moments of all the external forces acting on the portion of the beam to the left of the section; the origin of moments being taken in the section.

The *resisting moment* at any section is the algebraic sum of the moments of the internal forces in the section acting on the portion of the beam to the left, and exerted by the portion to the right of the section; the origin of moments being the same as for bending moment.

The *stress moment* or *moment of internal stress* at any section consists of the two equal and opposite moments of the forces exerted across the section by the two portions of the beam upon each other.

Referring again to Fig. 43, let us analyze further the forces in the section at C. Applying to the body AC the second condition of equilibrium ((2) of Art. 58), and taking an origin at a point in the section, we see that the algebraic sum of the moments of all the external forces acting on the beam to the left of the section plus the sum of the moments of the internal forces acting on AC in the section must equal zero. The former sum is defined as the *bending moment* at the given section. Represent it by M. The latter sum is defined as the *resisting moment* at the section, and must be equal to $-M$, by the above principle.

We have thus far referred only to the internal forces exerted by CB upon AC; but evidently the equal and opposite forces exerted by AC upon CB have a moment numerically equal to M. The equal and opposite moments of the equal and opposite forces of the stress in the section together constitute the *stress moment* in the section.

If the external forces applied to the beam are all vertical, the value of M will be the same at whatever point of the section the origin is taken; since the arm of each force will be the same for all origins in the same vertical line. If the loads and reactions are not all vertical, the value of M will generally depend upon what point in the section is taken as the origin of moments.

117. Curves of Shear and Bending Moment. — The *curve of shear* for a beam is a curve whose abscissas are parallel to the axis of the beam, and whose ordinate at any point represents the external shear at the corresponding section of the beam.

Let AB (Fig. 44) represent a beam loaded in any manner, and let $A'B'$ be taken parallel to AB. At every point of $A'B'$ suppose an ordinate drawn whose length shall represent the external shear at the corresponding section of AB. The line ab, joining the extremities of all these ordinates, is the *shear curve*. Positive values of the shear may be represented by ordinates drawn upward from $A'B'$, and negative values by ordinates drawn downward.

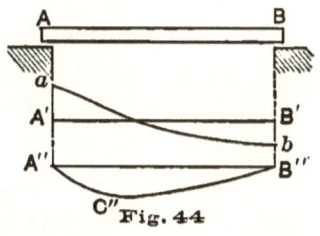

Fig. 44.

(Instead of drawing $A'B'$ parallel to the beam, it may be any other straight line whose extremities are in vertical lines through A and B.)

The *curve of bending moment* for a beam is a curve whose abscissas are parallel to the axis of the beam, and whose ordinate at any point represents the bending moment at the corresponding section of the beam. Thus in Fig. 44, $A''C''B''$ may represent the bending moment curve for the beam AB. (Evidently $A''B''$ might be inclined to the direction of AB, without destroying the meaning of the curve.) Positive and negative values of the bending moment will be distinguished by drawing the ordinates representing the former upward and those representing the latter downward from the line of reference $A''B''$.

118. Moment Curve a Funicular Polygon. — If the loads and reactions upon the beam are all vertical, every funicular polygon for these forces taken consecutively along the beam, is a curve of bending moments. Thus, let MN (Fig. 45) represent a beam under vertical loads, supported at the ends by vertical reactions. Draw a funicular polygon for the loads and reactions as shown.

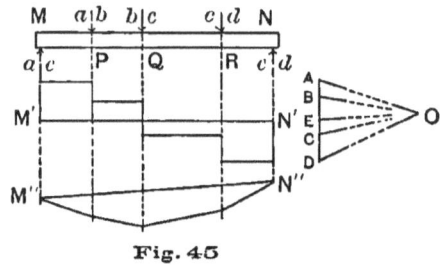

Fig. 45

Now, by definition, the bending moment at any section is equal to the moment of the resultant of all external forces acting on the beam to the left of the section, the origin of moments being taken in the section. By Art. 56, this moment can be found by drawing through the section a vertical line and finding the distance intercepted on it by the two strings corresponding to the resultant mentioned; the product of this intercept by the pole distance is the required moment. Hence, if oe (Fig. 45) is taken as axis of abscissas, the broken line made up of oa, ob, oc, od, is a "curve of bending moments."

119. **Design of Beams.** — The principles involved in the design of beams will not be here fully discussed. Every problem in design involves the determination of shears and bending moments throughout the beam; and the graphic methods of determining these will alone be considered in the following articles.

§ 2. *Beam Sustaining Fixed Loads.*

120. **Shear Curve for Beam Supported at Ends.** — Let MN (Fig. 45) represent a beam supported at the ends and sustaining loads as shown. Draw the force polygon $ABCD$, and with pole O draw the funicular polygon. The closing line is oe

(marked also $M''N''$), and OE drawn parallel to $M''N''$ fixes E, thus determining DE, EA, the reactions at the supports.

Take $M'N'$ as the axis of abscissas for the shear curve.

The shear at every section can be at once taken from the force polygon. For, remembering the definition of external shear (Art. 115) we have:

Shear at any section between M and P is EA (positive).
Shear at any section between P and Q is EB (positive).
Shear at any section between Q and R is EC (negative).
Shear at any section between R and N is ED (negative).

Hence the shear curve is the broken line drawn in the figure.

121. Moment Curve for Beam Supported at Ends. — As in Art. 118, it is seen that the funicular polygon already drawn (Fig. 45) is a bending moment curve for the given forces. For the bending moment at any section of the beam is equal to the corresponding ordinate of this polygon, multiplied by the pole distance.

For simplicity, it will be well always to choose the pole so that the pole distance represents some simple number of force-units.

The sign of the bending moment is readily seen to be negative everywhere, according to the convention already adopted (Art. 47).

122. Shear and Moment Curves for Overhanging Beam. — Consider a beam such as shown in Fig. 46, supported at Q and T, and sustaining loads at M, P, R, S, and N.

This case may be treated just as the preceding, care being exercised to take all the external forces (loads and reactions) in order around the beam.

The construction is shown in Fig. 46. First, the reactions at Q and T are found as in the preceding case, by drawing the funicular polygon, finding the closing line, and drawing OG parallel to it. The reactions are FG and GA. Then the value of the external shear can be found for any section from the definition, and is always given in magnitude and sign by a certain portion of the force polygon $ABCDEFGA$. The resulting curve is shown in the figure, $M'N'$ being the line of reference.

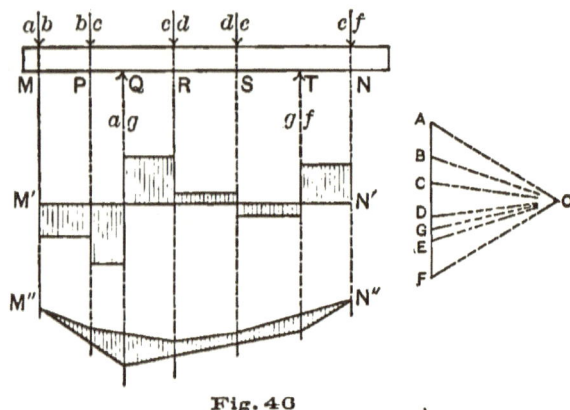

Fig. 40

Similarly, from the definition of bending moment, and the principle of Art. 56, the bending moment at any section is equal to the ordinate of the funicular polygon multiplied by the pole distance. The polygon is shaded in the figure to indicate the ordinates in question. It is seen that the bending moment is positive at every section of the beam.

123. **Distributed Loads.** — In all the cases thus far discussed, the loads applied to the beam have been considered as concentrated at a finite number of points. That is, it has been assumed that a finite load is applied to the beam *at a point.* Such a condition cannot strictly be realized, every load being in fact distributed along a small length of the beam. If the length along which a load is distributed is very small, no important error results from considering it as applied at a point.

In certain cases a beam may have to sustain a load which is distributed over a considerable part of its length, or over the whole length. Such a load may be treated graphically with sufficient correctness by dividing the length into parts, and assuming the whole load on any part to be concentrated at a point. The smaller these parts, the more nearly correct will the results be.

It may be remarked by way of comparison that while algebraic methods are most readily applicable to the case of distributed loads, the reverse is true of graphic methods, which are most easily applied in the very cases in which analytic methods become most perplexing.

124. **Design of Beam with Fixed Loads.** — The above examples are sufficient to explain the method of treating any simple beam under fixed loads. Under such loading the shear and bending moment at each section of the beam remain unchanged in value, and no further discussion is necessary as a preliminary to the design of the beam. We proceed next to the case of beams with moving loads.

§ 3. *Beam Sustaining Moving Loads.*

125. **Curves of Maximum Shear and Moment.** — When a beam sustains moving loads, the shear and moment at any section do not remain constant for all positions of the loads. In such a case it is the greatest shear or moment in each section that is to be used in designing the beam.

A *curve of maximum shear* is a curve of which the ordinate at each point represents the greatest possible value of the shear at the corresponding section of the beam for any position of the loads.

A *curve of maximum moment* is a curve whose ordinate at each point represents the greatest possible value of the bending moment at the corresponding section of the beam for any position of the loads.

In the following articles will be explained a method of determining any number of points of the curves just defined, in the case of a simple beam supported at the ends. The moving load will be taken to consist of a series of concentrated loads with lines of actions at fixed distances apart. An example of such a load is the weight of a locomotive and train; the points of application of the loads being always under the several wheels.

On Pl. III is represented a load-series consisting of two locomotives followed by a train whose weight is assumed to be uniformly distributed. The numbers given represent half the total weight of the train, being the weight borne by each of the two beams or trusses which sustain a single-track railroad.

126. Position of Loads causing Greatest Shear at a Given Section. — In order to determine the position of a given series of loads which causes the greatest positive shear at a given section of a beam, consider first the way in which the shear due to a single load varies as the position of the load changes.

Let A_1B_1 (Pl. III) represent the beam, and F_1 the section under consideration. A load in any position between F_1 and B_1 causes at F_1 a positive shear equal in magnitude to the reaction at A_1; a load between A_1 and F_1 causes at F_1 a negative shear equal to the reaction at B_1.

Let $A_1B_1 = l$, $A_1F_1 = l_1$, $F_1B_1 = l_2$. A load P on A_1F_1, distant x from A_1, causes at F_1 a shear $-\frac{Px}{l}$; as the load moves from A_1 to F_1 this shear varies from 0 to $-\frac{Pl_1}{l}$. A load P on F_1B_1, distant x from B_1, causes at F_1 a shear $\frac{Px}{l}$; as the load moves from B_1 to F_1 this shear varies from 0 to $\frac{Pl_2}{l}$. These results may be represented graphically as follows:

Influence line. — From a given reference line A_2B_2 (Pl. III (C)), let ordinates be erected such that the ordinate at any point J_2 represents (to a convenient scale) the shear at F_1 due to a unit load at J_1. The line joining the extremities of these ordinates is called an "influence line" for the shear in question.

The above values of the shear due to a load P in any position show that for the portion of the beam A_1F_1 the influence line is a straight line A_2F_3', F_2F_3' being equal in magnitude to $\frac{l_1}{l}$, and being laid off negatively; while for the portion F_1B_1 the influence line is the straight line B_2F_3, F_2F_3 being equal to $\frac{l_2}{l}$, and being laid off positively. The two lines A_2F_3' and F_3B_2 are seen to be parallel.

The influence line shows at a glance the effect of a load in any position in producing shear at F_1. Thus, a unit load at J_1 causes a positive shear equal to J_2J_3, and a load P at J_1 causes a positive shear equal to $P \times J_2J_3$.

The resultant shear at F_1 due to any number of loads may be found by multiplying each load by the corresponding ordinate of the influence line and taking the algebraic sum of the products.

Consider, now, the effect of any series of loads brought on the beam from the right. Every load causes positive shear at F_1 until the foremost load reaches the section, and the effect of each load increases as it approaches F_1. As soon as any load has passed the section, its effect is to cause negative shear, this effect decreasing as the load recedes from the section.

It thus appears that, in order that the shear at a given section may have its greatest positive value, (*a*) there should be little or no load to the left of the section, (*b*) the portion of the beam to the right of the section should be fully loaded, and (*c*) the loads at the right should be as near the section as possible.

These rules are not sufficiently definite to serve as an exact guide. Suppose the first load to be just at the right of F_1, and consider the effect of advancing the whole series of loads. As the first load passes the section the shear diminishes by an amount equal to that load. But as the loads advance the effect of every load in producing reaction at A_1 increases (thus increasing the shear at F_1), while no further *decrease* occurs until the second load passes the section. The net result of bringing the second load up to the section may or may not be to increase the shear over the value it has when the first load is just at the right of F_1. It is thus uncertain, without computation, whether the first load or some succeeding load should be brought up to the section in order to cause the greatest positive shear.

Instead of resorting to trials to determine for which position the shear is greatest, we may apply a simple rule which will now be deduced.

Let P_1 = magnitude of foremost load, P_2 = magnitude of second load, etc., W being the total load on the beam. Let l = total span, and m = distance between P_1 and P_2. Let x = distance from B_1 to the center of gravity of W when P_1 is at the given section. Let us compute the shear when P_1 is just at the right of the section, and then determine the effect of moving all loads to the left, until the second load comes to the section.

When P_1 is at the right of the section, the shear is equal to the reaction at A_1, say R. Then (calling V the external shear) we have

$$V = R = \frac{Wx}{l}.$$

If now the loads be moved until P_2 comes to a point just at the right of the section, the reaction due to W becomes $W\frac{(x+m)}{l}$, and the shear at the section becomes

$$V = R - P_1 = \frac{W(x+m)}{l} - P_1.$$

The increase of the shear is, therefore,

$$\frac{W(x+m)}{l} - P_1 - \frac{Wx}{l} = \frac{Wm}{l} - P_1.$$

This increase is plus or minus according as $\frac{Wm}{l}$ is greater or less than P_1; that is, as W is greater or less than $\frac{P_1 l}{m}$. Hence the following rule:

BEAM SUSTAINING MOVING LOADS. 105

The maximum positive shear in any section of the beam occurs when the foremost load is infinitely near the section, provided W is not greater than $\frac{P_1 l}{m}$. If W is greater than $\frac{P_1 l}{m}$, the greatest shear will occur when some succeeding load is at the section.

In the above discussion it has been assumed that in bringing P_2 up to the section no additional loads are brought upon the beam. If this assumption is not true, let W' be the new load brought on the beam, and x' the distance of its center of gravity from the right support when P_2 is at the section. Then the shear corresponding to this position is

$$V = R - P_1 = \frac{W(x+m)}{l} + \frac{W'x'}{l} - P_1;$$

and the increase of shear due to the assumed change in the position of the load is

$$\frac{Wm}{l} + \frac{W'x'}{l} - P_1.$$

Hence, in the statement of the above rule, we have only to substitute $Wm + W'x'$ for Wm; or, $W + \frac{W'x'}{m}$ for W. The additional load W' may be neglected except when the condition $W = \frac{P_1 l}{m}$ is nearly satisfied; for the term $W'x'$ will always be small.

The application of this rule is very easy, and will save much labor in the graphic construction of the shear curve.

If it is found that the first load should be past the section for the position of greatest shear, we may determine whether the second or the third should be at the section by an exactly similar method. We have only to apply the above rule, substituting P_2 for P_1, and for m the distance between P_2 and P_3.

127. Determination of Greatest Shear. — Having determined what position of the load-series causes the greatest shear at a given section, the value of this shear may be determined by the method already explained in the discussion of fixed loads (Art. 120).

Draw in succession the lines of action of the loads at their proper distances apart (1-2, 2-3, ..., Pl. III). Draw the force polygon * 1-2-3- ..., choose a pole O, and construct a funicular polygon for the series of loads. The same lines of action may be used for all positions of the load-series, for instead of moving the loads in the desired direction, the loads may be regarded as fixed and the beam moved in the opposite direction.

At (S), Pl. III, is shown a beam A_1B_1, of 64 ft. span, in such position that the shear at the section F_1, 16 ft. from A_1, has its greatest possible value. The second load is at the section instead of the first, for applying the test we have $P_1 = 8000$ lbs., $l = 64$ ft., $m = 8.1$ ft., $\frac{P_1 l}{m} = 63200$ lbs.; while the total load on the beam is 104,000 lbs. when the first load is at the section.

The beam being in the position shown at (S), vertical lines drawn through A_1 and B_1 intersecting the strings o1 and o2' at A and B, determine AB, the closing string of the funicular polygon for the system of forces acting upon the beam in the assumed position. If OK is the ray drawn parallel to AB, the reactions at A_1 and B_1 are $K1$ and $2'K$ respectively. The load 2-3 being just at the right of the section F_1, the shear at that section is equal to the reaction at A_1 minus the load 1-2; its value is therefore $K2$ in the force polygon.

The diagram of maximum positive shear is shown at (A), Pl. III, the ordinates representing greatest shear due to the moving loads being laid off upward from A_1B_1. Thus the value just determined is laid off from the point F_1. The entire curve is shown at (A), but the construction is given only for the section F_1.

The foregoing construction has referred only to moving loads. The actual greatest shear at any section is found by combining

---

* The lines representing the loads are for convenience drawn upward instead of downward. This does not, of course, affect the construction.

BEAM SUSTAINING MOVING LOADS. 107

the maximum live-load shear with the shear due to permanent loads. Shears due to the dead loads are represented at (A), Pl. III, being determined as follows.

128. Shear Curve for Combined Fixed and Moving Loads. — Let the beam sustain a fixed load of 25000 lbs. uniformly distributed along the beam. The shear close to the left support due to this load is equal to the reaction, or 12500 lbs.; and decreases as we pass to the right by $\frac{25000}{64}$ lbs. for each foot. At the middle of the beam the shear is zero, and at the right support it is $-$ 12500 lbs. Hence the shear curve is a straight line, and may be drawn as follows: From A_1 lay off an ordinate downward representing 12500 lbs., and from B_1 an ordinate upward representing 12500 lbs.; the straight line joining the extremities of these ordinates is the shear curve for the fixed load. Positive shears are laid off downward and negative shears upward for the reason that, if this be done, the greatest positive shear at any point due to fixed and moving loads is represented by the total ordinate measured between the shear curves for fixed loads and for moving loads. It is seen that at a certain point somewhere to the right of the center of the beam this greatest shear is zero, and for all sections to the right of this point it is negative. This point is determined by the intersection of the two shear curves.

129. Approximate Position of Loads causing Greatest Bending Moment at a Given Section. — Let it be required to determine the greatest bending moment at the section F_1 of the beam A_1B_1, shown on Pl. III. Let $A_1F_1 = l_1$, $F_1B_1 = l_2$, $A_1B_1 = l_1 + l_2 = l$. Consider the effect of a single load in any position.

The bending moment at F_1 due to a load P on A_1F_1, distant x from A_1, is $\frac{Pl_2x}{l}$. For a load P on F_1B_1, distant x from B_1, the bending moment at F_1 is $\frac{Pl_1x}{l}$. The variation of each of these values as the position of the load changes may be represented graphically as follows:

Influence line. — From A_2B_2 ((D), Pl. III), let ordinates be erected such that the length of the ordinate at any point represents, on a convenient scale, the bending moment at F_1, due to a unit load at the corresponding point of the beam. The line joining the extremities of these ordinates is the "influence line" for bending moments at F_1.

At a distance x from A_2, the ordinate has the value $\frac{l_2 x}{l}$, which varies from 0 at A_2 to $\frac{l_1 l_2}{l}$ at F_2. At a distance x from B_2, the value of the ordinate is $\frac{l_1 x}{l}$, which varies from 0 at B_2 to $\frac{l_1 l_2}{l}$ at F_2. The influence line therefore consists of the two straight portions A_2F_3, F_3B_2, where $F_2F_3 = \frac{l_1 l_2}{l}$. The ordinates are laid off downward, the bending moment being always negative according to the convention already adopted (Art. 47 and Art. 116).

The influence line having been drawn, the bending moment at F_1 due to a load at any point J_1 may be found by multiplying the load by the corresponding ordinate J_2J_3. The bending moment at F_1 due to the combined action of several loads is found by multiplying each load by the corresponding ordinate of the influence line and adding the products.

Since the sign of the bending moment due to a load is the same, whatever its position, and since the bending moment due to a given load increases as the load approaches the section, the following general principle may be stated:

The bending moment at any section has its greatest value when the beam is as fully loaded as possible, with the heaviest loads near the section.

This principle serves only as a rough general guide. An exact rule will be deduced in the following Articles.

130. **Load Polygon.** — Let a curve or broken line be drawn, of which the abscissa is horizontal, and the ordinate at any point represents (on the assumed force scale) the total load up

to that point, measured from some fixed point in the load series Such a curve or broken line may be called a *load polygon*. In the figure on Pl. IV, $Q'R'$ is a load polygon for the series of train loads shown on Pl. III, QR being a funicular polygon for the same series.

A simple relation exists between the load polygon and the funicular polygon, which is of use in the discussion of the problem now before us.

Let x and y be the coördinates of any point of the funicular polygon, the y-axis (LY) being vertical, and the x-axis any horizontal line LX. Let x', y' be the coördinates of the corresponding point of the load polygon, referred to a vertical axis ($L'Y'$), and a horizontal axis ($L'X'$) passing through the pole of the force polygon used in the construction of the funicular polygon.

Consider any point of the funicular polygon, for example a point on the string o2'. The value of $\frac{dy}{dx}$ at this point is the tangent of the angle between o2' and the horizontal. But since o2' is parallel to the corresponding ray of the force polygon, this tangent is equal to $\frac{y'}{H}$, where H is the pole distance used in drawing the funicular polygon. That is,

$$\frac{dy}{dx} = \frac{y'}{H},$$

or
$$y' = H\frac{dy}{dx} \quad \ldots \ldots \ldots \ldots (1)$$

131. Variation of Bending Moment at Any Section of a Beam. — Let A_1B_1 (at (A), Pl. IV) represent one position of a beam of span l, and let F_1 be any section at which it is desired to study the variation of the bending moment. Let $l_1 = A_1F_1$, $l_2 = F_1B_1$, and $z =$ horizontal distance of A_1 from the vertical axis $L'Y'$. Through A_1, B_1, and F_1 draw vertical lines, intersecting the funicular polygon in A, B, F, and the load polygon in A', B', F'. Let the vertical through F_1 intersect AB in G, and $A'B'$ in G'. Let the ordinates of A, B, F, G, measured from LX, be denoted by a, b, f, g; and the ordinates of A', B', F', G', measured from $L'X'$, by a', b', f', g'.

If M denotes the bending moment at F_1, and H the pole distance,
$$M = H \times FG = H(g-f); \qquad \ldots \ldots (2)$$

and it will now be shown that the rate of change of this bending moment (*i.e.* the change per unit of horizontal displacement) as the beam moves horizontally while the loads remain stationary is equal to $F'G'$ or $g'-f'$.

Since AGB is a straight line, and $\dfrac{AG}{GB} = \dfrac{l_1}{l_2}$, it follows that

$$\frac{g-a}{b-g} = \frac{l_1}{l_2},$$

or
$$g = \frac{l_2}{l}a + \frac{l_1}{l}b;$$

$\therefore$
$$FG = g - f = \frac{l_2}{l}a + \frac{l_1}{l}b - f. \qquad \ldots \ldots (3)$$

In an exactly similar manner it may be shown that

$$F'G' = g' - f' = \frac{l_2}{l}a' + \frac{l_1}{l}b' - f'. \qquad \ldots \ldots (4)$$

From (2) and (3),
$$M = H(g-f) = H\left(\frac{l_2}{l}a + \frac{l_1}{l}b - f\right) \qquad \ldots \ldots (5)$$

If the beam moves horizontally, the abscissas of A_1, B_1, and F_1 change at the same rate. Differentiating (5) with respect to z,

$$\frac{dM}{dz} = H\left(\frac{l_2}{l}\frac{da}{dz} + \frac{l_1}{l}\frac{db}{dz} - \frac{df}{dz}\right).$$

Now $\dfrac{da}{dz}$, $\dfrac{db}{dz}$, and $\dfrac{df}{dz}$ are equal to the values of $\dfrac{dy}{dx}$ at A, B, and F, respectively; therefore, from equation (1) of Art. 130,

$$a' = H\frac{da}{dz}, \quad b' = H\frac{db}{dz}, \quad f' = H\frac{df}{dz};$$

BEAM SUSTAINING MOVING LOADS.

hence

$$\frac{dM}{dz} = \frac{l_2}{l}a' + \frac{l_1}{l}b' - f';$$

or, comparing with equation (4),

$$\frac{dM}{dz} = g' - f' = F'G'. \quad \ldots \ldots \ldots \ldots (6)$$

132. Condition for Maximum or Minimum Bending Moment at Any Section. — As the beam moves relatively to the loads, FG (proportional to M) in general varies. As it passes through a maximum or a minimum value, $\frac{dM}{dz}$ must equal zero; but, from the relation just proved, this makes $F'G' = 0$. That is,

When the bending moment at any given section of the beam is a maximum or a minimum, the line $A'B'$ intersects the load polygon at a point directly opposite the section considered.

This condition may be stated in algebraic form as follows:

Let W_1 denote the total load between A_1 and F_1, W_2 the load between F_1 and B_1, and W the total load on the beam. Evidently,

$$W_1 = f' - a'; \quad W_2 = b' - f'; \quad W = b' - a'.$$

If the points F' and G' coincide, $f' - a'$ and $b' - f'$ are proportional to $A'G'$ and $G'B'$, that is to l_1 and l_2, and therefore,

$$\frac{W_1}{l_1} = \frac{W_2}{l_2} = \frac{W}{l}.$$

That is, when the bending moment at any section of the beam is a maximum or a minimum, the average load per unit length on each segment is equal to the average load per unit length on the whole span.*

---

* It is to be noted that if, at any point, there is a load which is concentrated in the strict mathematical sense (*i.e.*, a finite load applied at a mathematical point), the value of $\frac{dy}{dx}$ suffers a sudden change of value at the point of application of that load, and the equation

$$y' = H\frac{dy}{dx}.$$

does not hold at that point. This consideration does not, however, invalidate the above conclusions. For (*a*) a concentrated load in the strict mathematical sense does not occur

133. Discrimination between Maxima and Minima. — The magnitude of the bending moment is

$$M = H(g-f);$$

it is obvious from the form of the funicular polygon that g is always greater than f.

If $g'-f'$ is positive, $g-f$ (and therefore M) increases as z increases (*i.e.*, as the beam is displaced in the positive direction). If $g'-f'$ is negative, M decreases as z increases. Therefore, as M passes through a maximum value, the sign of $g'-f'$ changes from plus to minus; and as M passes through a minimum value, $g'-f'$ changes from minus to plus (the displacement of the beam being in the positive direction, that is, toward the right as the figure is drawn on Pl. IV).

If the load series consists of concentrated loads, $g'-f'$ will in general pass through zero only when a load is at one of the three points A_1, B_1, F_1. By applying the test just given, it may be seen that a maximum value can result only when a load passes F_1.

In applying the criterion for a position of maximum or minimum value of the bending moment, and in distinguishing positions causing maximum stress from those causing minimum values, it is not necessary to draw the line $A'B'$ for every trial position of the beam. A movable strip of paper representing the beam may be laid in any desired position, the points A' and B' approximately located by inspection, and a thread stretched between them. If the thread crosses the load polygon at a point directly opposite the section in question, the condition for

---

in practice, and (*b*) the ideal case of a finite load applied at a mathematical point may be treated as the limit of a case of distributed loading, so that even in that case the conclusions above derived hold, with slight change in the form of statement.

The series of concentrated loads actually coming upon a bridge in practice gives a load polygon which, while approaching the form shown on Pl. IV, differs from it in that the portions under the loads are not strictly vertical lines, but are lines of variable (very steep) inclination.

In applying the condition just deduced, a load at either of the three points A_1, B_1, F_1, is to be regarded as distributed in an arbitrary way along the small element of horizontal distance near the line used as its line of action.

maximum or minimum is fulfilled. Only positions which bring a load at the section need be tried. The accurate location of A' and B' is usually unnecessary.*

At (B), Pl. IV, is shown the beam A_1B_1 in such a position that $F'G' = 0$. It is seen that this position corresponds to a maximum value of the bending moment at F_1; for a displacement of the beam toward the right (positive) makes $g'-f'$ negative, while a displacement toward the left makes $g'-f'$ positive.

134. Determination of Bending Moment. — Having determined the position of the load series causing greatest bending moment at a given section, the value of the bending moment may be computed by the method already explained in the case of fixed loads. At (M), Pl. III, is represented the position of the beam A_1B_1, which causes the greatest bending moment at the section F_1, distant 16 ft. from A_1, the span being 64 ft. That the condition for a maximum is satisfied in this position may easily be verified. The total load on the beam is 112000 lbs., one-fourth of which is 28000 lbs. By regarding 5000 lbs. of the load 3–4 as belonging to the segment A_1F_1, the load on this segment is 28000 lbs. It is seen also that the beam is fully loaded, and that the heaviest loads are near the section. Moreover, the condition for maximum is not satisfied for any other position near the one chosen.

The closing string of the funicular polygon for all external forces acting on the beam in the position (M) is AB. The bending moment at F_1 is the product of the ordinate FG by the pole distance.

At (B), Pl. III, is shown the diagram of maximum bending moments for all sections of the beam. From A_1B_1 are laid off downward ordinates proportional to the maximum bending moments. The length FG just determined is laid off from F_1; this must be multiplied by the pole distance (100000 lbs. in the

---

* See paper by Professor Henry T. Eddy, Trans. Am. Soc. C. E., Vol. XXII.

diagram). The complete curve is drawn, but the construction for determining values of the maximum bending moment is shown only for the section F_1.

135. Moment Curve for Fixed Loads. — The greatest bending moment due to moving loads must be combined with the bending moment due to fixed loads. If the fixed load is uniformly distributed, as already assumed in the computation of shear, it may be divided into parts, each assumed concentrated at its center of gravity, and a funicular polygon drawn, using the same pole distance employed in the force diagram for moving loads. The ordinates of this funicular polygon may be laid off upward from the line A_1B_1, and their ends joined to form a continuous curve. The total ordinate from this curve to that already drawn for moving loads represents the true greatest bending moment at the corresponding section of the beam. The curve is shown at (B), Pl. III, but the construction is omitted, since it involves no new principle.

It is to be remembered that the bending moment found for any section is a possible value for the other section equally distant from the center of the beam, since the train may be headed in the opposite direction, and the same construction made, viewing the beam from the other side. The same statement holds as to shears.

136. Design of Beam sustaining Moving Loads. — In designing a beam to sustain moving loads, the greatest shear and bending moment that can come upon it for any position of the loads must be known for every section. The methods above given are sufficient for the determination of these quantities; and the problem of designing the beam will not be here further discussed.

CHAPTER VII. TRUSSES SUSTAINING MOVING LOADS.

§ 1. *Bridge Loads.*

137. General Statement. — The most important class of trusses sustaining moving loads is that of bridge trusses. The two main classes of bridges are highway bridges and railway bridges. The forms of trusses most commonly used differ for the two classes, as do also the amount and distribution of loads.

Before the design can be correctly made, the weights of the trusses themselves must be known. Since these weights depend upon the dimensions of the truss members, they cannot be known with certainty until the design is completed. The remarks made in Art. 82 regarding the design of roof trusses are here applicable.

In the following articles we shall give data available for preliminary estimates of truss weights.

Railway bridges of short span are frequently supported by rolled or built beams. When the span is longer than 100 ft. trusses should be used. (Cooper's "Specifications for Iron and Steel Railroad Bridges.")

138. Loads on Highway Bridges. — (1) *Permanent load.* — The permanent load sustained by a highway bridge truss includes the weight of the truss itself, of the lateral or "sway" bracing, of the floor and the beams and stringers supporting it. These weights are all subject to much variation, but, for purposes of preliminary design, the following formula, taken from Merriman's "Roofs and Bridges," may be used.

Let w = weight of bridge in pounds per linear foot; l = span in feet; b = width in feet. Then

$$w = 140 + 12\,b + 0.2\,bl - 0.4\,l.$$

(2) *Snow load.* — The weight of snow may be taken as in case of roof trusses (Art. 84). The values there given are probably in excess of those ordinarily employed in practice.

(3) *Wind load.* — The pressure of wind striking the bridge laterally is resisted by the chord members together with the lateral bracing. These constitute horizontal trusses, in which the stresses are to be found in the same way as for the main trusses of the bridge. As the determination of wind stresses requires the use of no special methods or principles, they will not be here considered. The student is referred to Burr's "Stresses in Roofs and Bridges," Merriman's "Roofs and Bridges," and other available works for a complete discussion of wind pressure and its effects on bridge trusses.

(4) *Moving load.* — The most dangerous moving load for a highway bridge is usually a crowd of people or a drove of animals. This is commonly taken as a uniformly distributed load, which may cover the whole bridge or any portion of it. Its value is variously taken at from 60 lbs. to 100 lbs. per square foot of area of floor, depending upon the span and upon local conditions.

It may be that in certain cases the greatest stresses will result from the passage of heavy pieces of machinery over the bridge, as, for example, a steam road roller. This should of course be considered in the design.

For a complete discussion of loads on highway bridges, the student is referred to Waddell's "Highway Bridges."

139. **Loads on Railway Bridges.** — (1) *Permanent load.* — The permanent load on a railway bridge includes (*a*) the weight of the track system, which is known or may be determined at the outset; (*b*) the weight of longitudinal stringers and cross-

beams, which can be determined before the trusses are designed; and (c) the weights of trusses. The weight of the track system may be taken at 400 lbs. per linear foot for a single track. (See Burr's "Stresses in Bridge and Roof Trusses"; Cooper's "Specifications for Iron and Steel Railroad Bridges"; Merriman's "Roofs and Bridges.") The total weight of track system and supporting beams and stringers varies from 450 lbs. to 600 lbs. per linear foot. (Merriman.) For spans less than 100 feet, Merriman gives the following formulas, in which w is the total dead load of the bridge in pounds per linear foot, and l is the span in feet:

For single track, $w = 560 + 5.6\,l$.

For double track, $w = 1070 + 10.7\,l$.

See also Art. 8 of Burr's work above cited.

(2) *Snow and wind.* — Railway bridges usually offer little opportunity for the accumulation of snow. Wind pressure is, however, an important factor. Besides the pressure upon the bridge itself, the pressure upon trains crossing the bridge must be considered. The latter is a moving load and may be dealt with in the same way as other moving loads. Its amount may be computed from the area of the exposed surface of the train and the known (or assumed) greatest pressure due to wind striking a vertical surface (Art. 85).

For further discussion of wind pressure, the student is referred to the works already cited. The graphic methods of determining stresses due to wind will be evident when the methods for vertical loads given in the following articles are understood.

(3) *Moving loads.* — The moving load to be supported by a bridge consists of the weights of trains. Such a load is applied to the track at a series of points, namely, the points of contact of the wheels. But the load is applied to the trusses only at the points at which the floor beams are supported. Hence the actual distribution of loads upon the truss is somewhat complex. It was formerly common to substitute for the actual

load a uniformly distributed load, thus simplifying the problem of determining stresses. It is now more usual to consider the actual distribution of loads for some standard type of locomotive used by the railroad concerned, or specified by its engineers. For examples of such distributions the student is referred to Cooper's "Specifications" already cited; also to Pl. III, and to the following portions of this chapter.

140. **Through and Deck Bridges.** — A bridge is called *through* or *deck* according as the floor system is supported at points of the lower or of the upper chord. In the former case, if the trusses are too low to require lateral bracing above, they are called *pony* trusses.

The weight of the truss itself is to be divided between the upper and lower joints. But the weight of the floor system and of the supporting beams and stringers comes wholly at the lower joints of a through bridge, or at the upper joints of a deck bridge. The moving load is, of course, applied at the same joints at which the floor system is supported.

If the floor system is supported directly upon the upper chord, as is sometimes the case, the moving load and part of the dead load produce bending in the chord members; the design is otherwise unaffected by this construction.

§ 2. *Truss regarded as a Beam.*

141. **Classification of Trusses.** — Since a bridge truss acts as a practically rigid body resting on supports at the ends or other points and sustaining vertical loads, it may be regarded as a beam, and trusses may be classified in the same way as beams (Art. 114). The only class to be here considered is that of *simple trusses*, or such as may be regarded as rigid bodies for the purpose of determining the reactions.

Cantilever trusses and continuous trusses are defined like the corresponding classes of beams (Art. 114). The most important case is that of a truss *simply supported* at the ends.

142. **External Shear for a Truss.** — If a truss be regarded as a beam, the external shear, resisting shear, and internal shearing stress at any section may be defined just as in Art. 115. In some forms of truss a knowledge of the external shear at any section makes it possible to compute readily the stresses in certain truss members. Thus, in the portion of a truss represented in Fig. 47, let the section MN cut three members, of which two are horizontal. (The member $x'y'$ is disregarded.) Since each member can exert forces only in the direction of its length, the external shear in the section MN must be wholly resisted by the diagonal xy; and the internal force in xy must be such that its resolved part in the vertical direction is equal in magnitude to the external shear in the section. Let the external shear V be represented by YZ (Fig. 47); draw YX parallel to yx and ZX horizontal; then XY will represent the stress in xy. If V is positive (Art. 115), the stress in xy is a tension. If V is negative, the stress is a compression. If the member xy were replaced by one sloping the other way from the vertical, these statements as to kind of stress would be reversed.

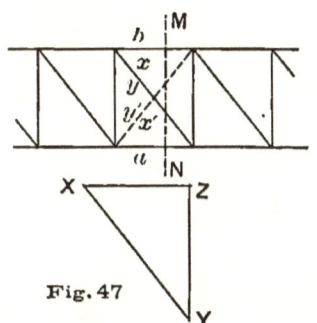

Fig. 47

If no two of the three members cut by any section are parallel, the stresses cannot be computed so simply, since all may contribute components of force to resist the external shear.

143. **Bending Moment for a Truss.** — In many cases the stresses in the truss members can be found from the values of the bending moment at different sections.

Thus, in Fig. 49, let a section MN be taken cutting three members as shown, and let the origin of moments be taken at the point of intersection of two of them (as bx and xy); then since the moments of the internal forces in these two will be zero, the resisting moment is equal to the moment of the internal force in the third member ay. The arm of this latter force is the perpendicular distance of its action line from the origin. Call it h, and let p represent the force itself, and M the bending moment. Then, numerically,

$$ph = M. \quad \therefore p = \frac{M}{h}.$$

If M is positive (in accordance with the convention of Art. 47), p must act from right to left, hence the stress in ay is a compression. If M is negative, the stress is a tension. (It must be remembered that the forces whose moments make up the bending moment M act upon the portion of the truss to the left of the section.)

§ 3. *Truss sustaining Any Series of Moving Loads.*

144. General Method of Determining Maximum Stresses in Truss Members. — When a truss carries a definite series of moving loads, the problem of determining the maximum stress in any member involves (1) the determination of the proper position of the loads and (2) the actual computation of the stress for a known position of the loads. The second part of this problem involves only the principles already explained in the discussion of trusses carrying fixed loads. The first part requires special consideration.

Although the treatment of trusses with parallel chords is in some respects simpler than that of the case of non-parallel chords, it is advantageous to consider the problem in as general a manner as possible. The following discussion will therefore refer to the general case of a truss of any form, subject only to the restriction that there are no redundant members, so that the truss may be completely divided by cutting three members, of which any chosen member may be one.* No restriction will be

---

* A case not included in this general description, while still offering a determinate problem, is that of a truss with subordinate bracing. This important case will be considered in a following discussion.

made as to the distribution of the loads. Trusses with parallel chords, and particular distributions of loading, may be treated as special cases of the general problem.*

145. Computation of Stress when Position of Loads is Known. — The general "method of sections" (Art. 69) may be applied to determine the stress in any member when the position of the load series is known.

Thus, referring to Pl. V, let 1–2, 2–3, ... represent the lines of action of the successive loads, 1–2–3– ... the force polygon for the loads, and let a funicular polygon be drawn as shown. The truss is represented in two positions marked (A) and (B) respectively. In connection with (A) is shown the construction for determining the stress in the member C_1D_1, and in connection with (B) the construction for determining the stress in C_1H. Referring to the former case, let a section be taken through the three members C_1D_1, C_1F_2, H_1F_2; then the internal forces in the three members cut are in equilibrium with the external forces acting on the portion of the truss on either side (as the left) of the section. These forces are the reaction at A_1 and all the loads from A_1 to C_1. Applying the principle of moments, the origin may be taken at F_2, the point of intersection of the lines C_1F_2 and H_1F_2, thus eliminating the forces acting in these lines.

The moment of any load is equal to the sum of the moments of any components which may replace it. Hence for any load between A_1 and C_1 (though actually coming upon the truss at the panel joints) the moment may be computed as if it were applied to the truss at a point in the vertical line through its actual point of application on the floor or track. But in the case of a load on the panel C_1D_1, only one component comes upon the portion of the truss to which the moment equation is to be applied; it is therefore necessary to determine the portion of every such load which is actually carried at C_1. If loads between C_1 and D_1 are carried wholly at the two points C_1 and

---

* The case of parallel chords, with graphic methods of treatment, receives very full discussion in a paper by Professor Henry T. Eddy, Trans. Am. Soc. C. E., Vol. XXII (1890).

D_1, the portion carried at each of these points is easily determined, as will be shown.*

From A_1 and B_1 draw vertical lines intersecting the funicular polygon at A and B respectively; the line AB is the closing line of the funicular polygon for the external forces acting upon the truss when the load series has the assumed position. On the panel C_1D_1 is the load $6'7'$. Treating C_1D_1 as a simple beam, the components replacing this load at C_1 and D_1 are found (Art. 45) as follows: Draw vertical lines through C_1 and D_1 intersecting the strings $o6'$ and $o7'$ in C and D; from the pole O draw the ray OK parallel to CD; the required components of $6'7'$ are represented in the force polygon by $6'K$ and $K7'$. The same construction might be made for every panel, and it is seen that the series of lines such as CD would form the funicular polygon for the loads as actually applied to the truss at the several joints. But the construction is here needed only for the panel C_1D_1, since a load on any other panel comes wholly upon one of the two portions into which the truss is divided by the section taken.

The sum of the moments of the external forces acting upon the left portion of the truss may be found graphically by the method of Art. 56. That is, through the origin of moments a line is drawn parallel to the resultant of the forces (a vertical line in this case); its intercept FG, between the strings AB and CD, multiplied by the pole distance, gives the required moment. The origin being taken at F_2, this moment is equal to the moment of the internal force in the member C_1D_1; hence that force may be found by dividing the moment by the perpendicular distance from F_2 to C_1D_1.

Referring next to (B) and considering the member C_1H, the stress in this member may be found in a similar manner. The

---

* It is to be noticed that the actual distribution of any load among the different joints is not determinate by simple methods, if the longitudinal beams supporting the floor system are supported at more than two points. It will here be assumed that the portion of such a beam between two supports acts as a simple beam. The results of this assumption are probably as reliable as could be obtained by a more elaborate discussion.

points A, B, C, D, F, G have the same meanings in this construction as in the preceding, but AB and CD must be produced in order to find the intercept FG, because the origin of moments is at F_1, beyond the end of the span. The moment of the internal force in C_1H is equal to FG multiplied by the pole distance; dividing this moment by the perpendicular distance from F_1 to C_1H gives the magnitude of the required force.

The kind of stress in any member may be determined by inspection of the conditions, as in Art. 111.

If the stress in a member is determined by the foregoing method, the steps of the process are the same, whatever member be under consideration. These steps are as follows:

(1) A section is taken completely dividing the truss and cutting only three members, one of which is the member whose stress is to be determined, and one a member of the chord carrying moving loads.

(2) Taking origin at the intersection of those two of the three members cut whose stresses are not in consideration, equate the sum of the moments of the external forces acting upon one portion of the truss to the moment of the internal force in the member considered.

146. **Definitions and Notation.** — The following definitions and notation will conduce to clearness and generality in the ensuing discussion.

Let the origin of moments, assumed in determining the stress in any member by the foregoing general method, be called the *moment-center* for that member.

Let the *stress-moment* for any member be defined as the moment of the stress in that member with respect to the moment-center just described. Let the sign of the stress-moment be specified in the following manner:

Rotation being called positive or negative in accordance with the convention adopted in Art. 47, let the stress-moment be called positive when it resists a tendency of the left portion of the truss to rotate negatively (and of the right portion to rotate positively).

In all cases let the left and right ends of the span be marked A_1 and B_1, and let C_1 and D_1 designate the left and right ends of the panel cut by the assumed section. Let the vertical line through the moment-center cut A_1B_1 (produced if necessary) in a point marked F_1.

Distances from left to right being positive, and from right to left negative, let the following notation be used:

$$A_1F_1 = l_1, \quad F_1B_1 = l_2, \quad A_1B_1 = l_1 + l_2 = l;$$
$$C_1F_1 = n_1, \quad F_1D_1 = n_2, \quad C_1D_1 = n_1 + n_2 = n.$$

The sign of each of the four quantities l_1, l_2, n_1, n_2 is to conform in all cases to the direction, as indicated by the order of the letters used in defining it. Thus $l_1 = A_1F_1$ is positive or negative, according as F_1 is to the right or to the left of A_1.

147. Effect of a Single Load. — *Influence line.* — A general idea of the position of a load series which will cause the greatest stress in any member of a truss may be obtained by considering the effect of a single load placed in any position. The details of the discussion will vary with the form of the truss, but the method of reasoning may be sufficiently illustrated by reference to the form shown at (A) or (B), Pl. V.

The relative effect of loads in different positions in producing stress in a given member may be clearly represented by means of an "influence diagram," analogous to the diagrams used in the discussion of shear and bending moment in a beam (Arts. 126 and 129).

From a given base line let ordinates be erected such that the ordinate at any point represents (to a convenient scale) the stress due to a unit load at the corresponding point of the span. The line joining the extremities of these ordinates is the *influence line* [*] for the member in question.

---

[*] For a discussion of influence lines, see paper by Professor George F. Swain, Trans. Am. Soc. C. E., Vol. XVII, p. 21 (1887).

It will be shown that there is a simple method, of complete generality, by which the influence line may be drawn for any member whatever. In order to explain this method clearly, it is necessary to adopt a convention as to signs.

In laying off ordinates of the influence line, let positive values be drawn upward and negative values downward from the reference line. *Let the sign of the ordinate agree in all cases with that of the stress-moment* (Art. 146). With this convention, a positive ordinate may correspond to either a compressive or a tensile stress, depending upon the relation of the moment-center to the member under consideration. From the sign of the stress-moment the kind of stress may always be readily determined by inspection of the truss diagram.

In Fig. 48 (Pl. VIII) are shown the forms assumed by the influence line for three typical members of the truss shown; (A) represents the case of the chord member C_1D_1, (B) that of the chord member $C_1'D_1'$, and (C) that of the web member $C_1'D_1$. The method of constructing these diagrams will now be explained.

Member of loaded chord. — Consider first the member C_1D_1. The moment-center is C_1', and F_1 falls between C_1 and D_1, the corresponding position of F_2 being shown at (A). Here l_1, l_2, n_1, n_2 are all positive (Art. 146). A load P between D_1 and B_1, distant x from B_1, causes a reaction $\frac{Px}{l}$ at A_1, and a stress-moment equal to $\frac{Pl_1x}{l}$. This varies directly as x, being 0 at B_1 and $\frac{Pl_1(l_2-n_2)}{l}$ at D_1. Similarly, the stress-moment due to a load P between A_1 and C_1, distant x from A_1, is $\frac{Pl_2x}{l}$, which varies from 0 at A_1 to $\frac{Pl_2(l_1-n_1)}{l}$ at C_1. If the moment-arm of the required stress is t, the ordinates of the influence line at C_2 and D_2 are

$$C_2C_3 = \frac{l_2(l_1-n_1)}{lt} \text{ and } D_2D_3 = \frac{l_1(l_2-n_2)}{lt};$$

and the straight lines A_2C_3 and D_3B_2 are the portions of the influence line for the corresponding portions of the span. A load P between C_1 and D_1, distant x from C_1, comes upon the truss partly at C_1 and partly at D_1; the former part is $\dfrac{P(n-x)}{n}$, the latter $\dfrac{Px}{n}$. The total stress-moment due to such a load is therefore

$$\frac{P(n-x)}{n} \cdot \frac{l_2(l_1-n_1)}{l} + \frac{Px}{n} \cdot \frac{l_1(l_2-n_2)}{l};$$

putting $P=1$ and dividing by t, this expression gives the ordinate of the influence line for any point between C_1 and D_1. The influence line for this portion of the span is therefore the straight line C_3D_3. It will be noticed that, if the straight line B_2D_3 be continued, its ordinate at F_2 will be $F_2f_2=\dfrac{l_1l_2}{lt}$; if A_2C_3 be continued, its ordinate at F_2 will have the same value.

Member of chord not carrying moving loads. — Next consider the member $C_1'D_1'$. The moment-center is D_1, F_1 falls at D_1, and F_2 coincides with D_2 as shown at (B), Fig. 48. If the reasoning used in the preceding case be repeated, it is seen that the values there given for the ordinates of the influence line apply also to this case. Since in the present case $n_2=0$, the value of the ordinate at D_2 reduces to $\dfrac{l_2l_1}{lt}$. Further, it is seen that $A_2C_3D_3$ is a straight line.

Web member. — For the member $C_1'D_1$ the moment-center is at the intersection of $C_1'D_1'$ and C_1D_1; F_1 falls to the left of A_1, the corresponding position of F_2 being shown at (C), Fig. 48. Here it will be noticed that l_1 and n_1 are negative. If, however, the method used in the preceding cases is again applied, it is found that the values above given for the stress-moment due to loads on D_1B_1 and A_1C_1 are correct for the present case both in magnitude and in sign. Thus, for a load P on D_1B_1, distant x from B_1, the reaction at A_1 is $\dfrac{Px}{l}$, and the stress-moment is $\dfrac{Pl_1x}{l}$. This stress-moment is negative (according to the convention adopted in Art. 146); but since l_1 is negative, the expression $\dfrac{Pl_1x}{l}$ is correct in magnitude and sign. For a load P on A_1C_1, distant x from A_1, the stress-moment is positive, and is given in magnitude and sign by the same expression as in the preceding cases, $\dfrac{Pl_2x}{l}$. The ordinates at C_2 and D_2 also have the general values

$$C_2C_3=\dfrac{l_2(l_1-n_1)}{lt},\ D_2D_3=\dfrac{l_1(l_2-n_2)}{lt};$$

the latter expression being negative (because l_1 is negative and l_2-n_2 positive) while the former is positive (because l_2 and l_1-n_1 are both positive). Further, if A_2C_3 and D_3B_2 be prolonged, they have a common ordinate $\dfrac{l_1l_2}{lt}$ at the point F_2. This ordinate is negative, agreeing in sign with l_1.

General result. — It is thus seen that the construction of the influence line may follow one general rule in all cases. This rule may be stated as follows, designating points by letters, as above:

From the point F_2 erect an ordinate of magnitude and sign $\dfrac{l_1l_2}{lt}$. From the extremity of this ordinate draw straight lines to A_2 and B_2, the former intersecting the vertical through C_2 in C_3, the latter intersecting the vertical through D_2 in D_3. The line $A_2C_3D_3B_2$ is the required influence line.*

148. Approximate Position of Load-Series causing Greatest Stress in a Given Member. — Inspection of the influence diagram for a given member enables the position of loads causing maximum stress to be estimated approximately.

Member of chord carrying moving loads. — For the member C_1D_1 (Fig. 48, Pl. VIII), the influence diagram (A) shows that

---

* The same general rule applies when F_1 falls between A_1 and C_1, or between D_1 and B_1, n_1 being negative in one case, and n_2 in the other, while both l_1 and l_2 are positive in both cases. These cases occur in no ordinary form of truss, except that with subordinate bracing. This form is discussed in Art. 160, and the influence diagrams for negative values of n_1 and of n_2 are shown on Pl. VIII.

all loads on the truss produce the same kind of stress, and that the effect of a given load is greater the nearer it is* to D_1. Therefore, with any given load series, it is to be expected that the greatest stress will occur when (a) the truss is heavily loaded throughout the span, and (b) the heaviest loads are near the point D_1.

Chord not carrying moving loads. — In a similar manner, the influence diagram (B) shows that the greatest stress in $C_1'D_1'$ is to be expected when (a) the entire span is heavily loaded, and (b) the heaviest loads are near D_1.

Web member. — Diagram (C), Fig. 48, Pl. VIII, shows that for loads between A_1 and E_1 the stress-moment is positive, while for loads between E_1 and B_1 it is negative. Hence for the maximum stress of the kind corresponding to a positive stress-moment (a) the portion A_1E_1 should be heavily loaded while little or no load is between E_1 and B_1, and (b) heavy loads should be near C_1. For maximum stress of the kind corresponding to a negative stress-moment, (a) E_1B_1 should be heavily loaded while little or no load is on A_1E_1, and (b) heavy loads should be † near D_1.

The above principles serve as an approximate general guide, but do not enable us to determine the position of loads for maximum stress *exactly*. A more definite discussion of the problem will now be given.

149. **Criterion for Position of Loads causing Maximum or Minimum Stress in Any Member of a Truss.** — Let a funicular polygon and a load polygon be drawn for the given series of loads (Pl. V).

As shown in Art. 130, if y denotes any ordinate of the funicular polygon, and y' the corresponding ordinate of the

---

* It may, of course, happen that C_2C_8 is greater than D_2D_8. In such case, the effect of a load is greater the nearer it is to C_1.

† The nature of the load-series may be such that a compromise must be made between requirements (a) and (b).

load polygon (the axes being LX, LY, and $L'X'$, $L'Y'$, respectively),

$$y' = H \frac{dy}{dx} \quad \ldots \ldots \ldots \ldots \quad (1)$$

Consider the stress in any member of the truss shown at (A), Pl. V, as C_1D_1. Taking a section through C_1D_1, C_1F_2, and H_1F_2, the origin of moments for computing the required stress (the moment-center) may be taken at F_2. Let vertical lines through A_1, B_1, C_1, D_1 intersect the funicular polygon in A, B, C, D, and the load polygon in A', B', C', D'. Through F_2 draw a vertical line intersecting AB in G and CD in F; then (Art. 145) the moment of the required stress with respect to the origin F_2 is

$$M = H \times FG,$$

where H is the pole distance used in drawing the funicular polygon. If the load series moves (or if the truss is considered as moving while the loads remain stationary), the stress in C_1D_1 is always proportional to FG, and when the stress is a maximum or a minimum, the intercept FG is also a maximum or a minimum.

Let the vertical through F_1 intersect $A'B'$ in G' and $C'D'$ in F'. It will now be shown that, as the truss moves horizontally, the change of FG (and therefore of M) per unit distance moved is always proportional to $F'G'$. The reasoning is similar to that employed in Art. 131.

Let a, b, c, d, f, g, denote the ordinates of A, B, C, D, F, G, respectively, measured from the horizontal axis LX; and a', b', c', d', f', g', the ordinates of A', B', C', D', F', G', measured from the axis $L'X'$. Also, let z denote the abscissa of the point A_1, measured from the axis $L'Y'$, and let l_1, l_2, l, n_1, n_2, n have the meanings assigned in Art. 146.

It may be shown as in Art. 131 that

$$g = \frac{l_2}{l}a + \frac{l_1}{l}b;$$

and by exactly similar reasoning,
$$f = \frac{n_2}{n}c + \frac{n_1}{n}d.$$

Therefore
$$M = H \times FG = H(g-f) = H\left(\frac{l_2}{l}a + \frac{l_1}{l}b - \frac{n_2}{n}c - \frac{n_1}{n}d\right). \quad (2)$$

Similar reasoning applied to the lines $A'B'$, $C'D'$ leads to the equation
$$F'G' = g' - f' = \frac{l_2}{l}a' + \frac{l_1}{l}b' - \frac{n_2}{n}c' - \frac{n_1}{n}d'. \quad \ldots \ldots (3)$$

The rate of change of M as the truss moves horizontally is found by differentiating (2) with respect to z. Observing (as in Art. 131) that $\frac{da}{dz}, \frac{db}{dz}, \ldots$ are the values of $\frac{dy}{dx}$ at $A, B, \ldots$, equation (1) shows that
$$H\frac{da}{dz} = a', \quad H\frac{db}{dz} = b', \ldots;$$

the differentiation of equation (2) therefore gives
$$\frac{dM}{dz} = \frac{l_2}{l}a' + \frac{l_1}{l}b' - \frac{n_2}{n}c' - \frac{n_1}{n}d'.$$

Comparing with (3),
$$\frac{dM}{dz} = g' - f' = F'G'. \quad \ldots \ldots \ldots \ldots (4)$$

It follows from equation (4) that when the position of the loads is such that M is a maximum or a minimum (that is, when the stress in C_1D_1 is a maximum or a minimum), the ordinate $F'G'$ is equal to zero.

The above reasoning is general,* and the conclusion applies to any member of the truss. In the case of chord members the point F_1 will in general fall within the span; only in exceptional

---

* That the above reasoning is rigorous for the case of concentrated loads as actually applied to the truss in practice follows from the considerations mentioned in the note in Art. 132. With slight change in the form of statement, the conclusion holds even in the ideal case of concentrated loads in the strict mathematical sense.

forms of truss will it fall without the panel C_1D_1, though it may coincide with C_1 or D_1. The figure for the case of a web member is shown at (B), Pl. V, the letters A, B, C, D, F, G, with or without subscripts or accents, denoting corresponding points in the constructions shown at (A) and at (B). The point F_1 generally falls without the span in case of a web member; it is possible that it should fall between A_1 and C_1 or between D_1 and B_1. In any case equation (4) holds, and the condition for maximum or minimum stress takes the above general form.

The criterion above reached ($F'G' = 0$) is easy of application. Let a strip of cardboard be prepared, showing the truss-skeleton and the moment-centers for all members of the truss. By applying this to the diagram showing the load polygon, the points A' and B' may be located approximately by inspection (an accurate determination of them not usually being necessary for the purpose in hand); by stretching a thread through A' and B', the point G' may be located approximately; then, holding one point of the thread at G', it may be stretched so as to pass through one of the points C' and D', and if it passes through the other also, the condition for maximum or minimum stress is satisfied. As will be shown presently, when the load-series consists of concentrated loads, a load will generally be at C_1 or D_1 when the stress is a maximum; the load polygon will therefore be vertical (or very steep) at C' or at D', so that one of these points may be shifted vertically without changing appreciably the position of the other or of either A' or B'. In some cases (but not usually) it may be necessary to locate A', B', and G' accurately.

150. **Positions giving Maximum and Minimum Values of the Stress Distinguished.** — If $g-f$ and $g'-f'$ have like signs, the magnitude of $g-f$ increases with a positive displacement, and decreases with a negative displacement; if they have unlike signs, $g-f$ decreases with a positive displacement, and increases with a negative displacement. It follows that, if the truss is in

a position of maximum stress for the member considered, a small positive displacement makes the signs of $g-f$ and $g'-f'$ unlike, while a small negative displacement makes their signs like; while the reverse is true for small displacements from a position of minimum stress. (The words "maximum" and "minimum" here refer to *magnitude*, irrespective of the nature of the stress.)

A complete discussion of this principle would require the consideration of various forms of truss. It will be sufficient to illustrate the application to the form shown on Plates V and VI.

151. Application of General Principles. — On Pl. VI is represented a truss of 60 ft. span, divided into six equal panels. The greatest depth of the truss is 12 ft.; the lower chord (carrying the moving loads at the joints) is straight, while the upper joints lie upon the arc of a circle. The two web members intersecting at any upper joint are of equal length and slope.

The moving load consists of the two locomotives and train represented on Pl. III, for which a load polygon and a funicular polygon are drawn on Pl. VI. The diagram represents half the actual weight of the train, it being assumed that the load is divided equally between two trusses.

The construction for determining the greatest stress in the chord member C_1D_1 is shown at (A), Pl. VI, and the construction for the case of the web member HD_1 is shown at (B). Like lettering is used in the two cases.

Member of lower chord. — From the approximate general rule deduced in Art. 148, it follows that, in order that the stress in C_1D_1 may have its greatest value, the truss must be loaded as completely as possible. Moreover, since the effect of a given load is greater the nearer it is to the panel C_1D_1, the positions first to be tested are those bringing the heaviest loads upon or near the panel.

TRUSS SUSTAINING ANY SERIES OF MOVING LOADS. 133

The condition for maximum or minimum stress is found to be satisfied in the position in which the truss is shown in the figure, the heavy load 5'6' being at D_1; but not in any other position in which the heaviest loads are upon or near the panel C_1D_1 and the truss fully loaded. The criterion of Art. 150 shows conclusively that this position gives a maximum and not a minimum value of the stress; for $g-f$ is positive, and a positive displacement of the truss makes $g'-f'$ negative, while a negative displacement makes it positive.*

The value of the maximum stress in C_1D_1 may now be determined. The pole distance is 100000 lbs.; $FG=9.9$ ft.; F_2F_1 (the moment-arm of the stress) is 12 ft. The required stress is therefore a tension of

$$\frac{9.9 \times 100000}{12} = 82500 \text{ lbs.}$$

Web member. — Referring to (*B*), Pl. VI, consider the member HD_1. From Art. 148 it is known that when this member sustains its greatest tension the portion D_1B_1 must be as completely loaded as possible (with A_1C_1 probably free from loads), while to produce the greatest compression A_1C_1 must be as fully loaded as possible (with D_1B_1 probably free from loads). Whether, in either case, loads should be between C_1 and D_1 is not known without applying an exact test.

Considering the case of tension, let the load series be brought from the right until the foremost load reaches the panel, and let the criterion for maximum stress be applied to successive positions until one is found in which it is satisfied. It is found that the criterion is satisfied when the second load (2–3) is at D_1. That this position gives a maximum and not a minimum value of the stress is found by applying the test of Art. 150. For $g-f$ is negative; a positive displacement of the truss

---

* It will be noticed that, with a series of concentrated loads, $F'G'$ will, as a rule, pass through zero only when a load passes one of the four points A_1, B_1, C_1, D_1. Further, when F' is between C' and D', a load at A_1 or B_1 (when $F'G'=0$) will correspond to a minimum stress, while a load at C_1 or D_1 will correspond to a maximum.

makes $g'-f'$ positive, while a negative displacement makes it negative.*

The value of the maximum tension in the member HD_1 may now be found by the method of moments. The intercept $FG = 8.8$ ft.; the pole distance is 100000 lbs.; the arm $F_1J = 39.6$ ft. The required tension is, therefore,†

$$\frac{8.8 \times 100000}{39.6} = 22200 \text{ lbs.}$$

152. Algebraic Statement of Condition for Maximum or Minimum Stress. — The general condition above deduced admits of simple algebraic expression. For this purpose let

$P_1 =$ total load on A_1C_1,
$P_2 =$ total load on D_1B_1,
$Q\ \ =$ total load on C_1D_1,
$W =$ total load on $A_1B_1 = P_1 + P_2 + Q$.

Let l_1, l_2, n_1, n_2, l, n have the meanings assigned in Art. 146, their algebraic signs being carefully observed.

Referring to Fig. (A), Pl. V, the point G' divides $A'B'$ into segments proportional to l_1 and l_2, and F' divides $C'D'$ into segments proportional to n_1 and n_2. Also, with the notation of Art. 149, it is seen that

$$c' - a' = P_1; \quad b' - d' = P_2;$$
$$d' - c' = Q; \quad b' - a' = W = P_1 + P_2 + Q.$$

Since $A'G'B'$ is a straight line,

$$\frac{g' - a'}{l_1} = \frac{b' - g'}{l_2} = \frac{b' - a'}{l} = \frac{W}{l};$$

---

* Here again it is seen that the condition for maximum or minimum stress will, in general, be satisfied only when a load is at one of the four points A_1, B_1, C_1, D_1. In this case a *maximum* stress of the kind now under consideration will occur only when a load is at D_1.

† These results, being obtained from small-scaled drawings, are not given as accurate values of the stresses, the present object being merely to illustrate the method of procedure. By careful work, and with drawings made to a suitable scale, a sufficient degree of accuracy may be obtained by the above method.

TRUSS SUSTAINING ANY SERIES OF MOVING LOADS. 135

hence
$$g' - a' = \frac{Wl_1}{l}, \dots \dots \dots \dots (1)$$

$$b' - g' = \frac{Wl_2}{l}. \dots \dots \dots \dots (2)$$

Similarly, since $C'F'D'$ is a straight line,

$$\frac{f' - c'}{n_1} = \frac{d' - f'}{n_2} = \frac{d' - c'}{n} = \frac{Q}{n};$$

hence
$$f' - c' = \frac{Qn_1}{n}; \dots \dots \dots \dots (3)$$

$$d' - f' = \frac{Qn_2}{n}. \dots \dots \dots \dots (4)$$

Subtracting equation (3) from equation (1), and (4) from (2), and writing P_1 and P_2 instead of $c' - a'$ and $b' - d'$, there result the equations *

$$g' - f' = \frac{Wl_1}{l} - \frac{Qn_1}{n} - P_1; \dots \dots \dots (5)$$

$$f' - g' = \frac{Wl_2}{l} - \frac{Qn_2}{n} - P_2. \dots \dots \dots (6)$$

If the points F' and G' coincide, as is the case when the condition for maximum or minimum stress is satisfied, $g' - f' = 0$, and equations (5) and (6) may be written †

$$\frac{P_1 + \frac{n_1}{n}Q}{l_1} = \frac{P_2 + \frac{n_2}{n}Q}{l_2} = \frac{W}{l} \dots \dots \dots (7)$$

Although the foregoing discussion has referred directly to the

---

* Equations (5) and (6) are not independent; either may be derived from the other by means of the relations $P_1 + P_2 + Q = W$, $l_1 + l_2 = l$, $n_1 + n_2 = n$.

† It is interesting to compare this result with the condition for maximum or minimum bending moment for a beam, given in Art. 132. If the portion of the load polygon from C' to D' were replaced by the straight line $C'D'$ (*i.e.* if the load Q were uniformly distributed over the panel), the total load from A_1 to F_1 would be $P_1 + \frac{n_1}{n}Q$, and the total load from F_1 to B_1 would be $P_2 + \frac{n_2}{n}Q$. The condition for greatest stress in C_1D_1 is therefore identical with the condition for maximum bending moment at F_1, on the assumption that the load Q is uniformly distributed. This interpretation of equation (7) is of use only when F_1 falls between C_1 and D_1, although it holds in a mathematical sense in all cases.

case in which F_1 falls between C_1 and D_1, so that l_1, l_2, n_1, n_2 are all positive, the result holds also when any of these quantities are negative, signs being duly observed in substituting their values in equation (7).

If the member considered belongs to that chord to which the moving loads are applied (as at (A), Pl. V), l_1 and l_2 will be positive, and one or both the quantities n_1 and n_2 will be positive. If one of the members C_1F_2, F_2D_1 were vertical, either n_1 or n_2 would be zero. In no truss of ordinary design, without subordinate bracing, would n_1 or n_2 be negative in case of a chord member.

In case of a member of the chord carrying no moving loads, the point F_1 falls at one end of the panel C_1D_1, and either n_1 or n_2 becomes zero. This may be seen by referring to the member H_1F_2, Pl. V, (A).

In case of a web member (as C_1H, Pl. V, (B)), F_1 falls without the panel C_1D_1 (making n_1 or n_2 negative); in the form of truss here figured F_1 also falls without * the span A_1B_1 (making l_1 or l_2 negative). In this case equation (7) may be put into more convenient form. Consider, for example, the maximum tension in the member D_1H, in the truss shown at (B), Pl. VI. From the general principles stated in Art. 148 it follows that loads on D_1B_1 cause tension in this member, while loads on A_1C_1 cause compression. If possible, therefore, equation (7) should be satisfied by a position making P_1 zero or small. The equation may be written

$$Q = \frac{n}{n_1}\left(\frac{l_1}{l}W - P_1\right); \quad \ldots \ldots \ldots \ldots (8)$$

and this is to be satisfied by a position making $P_1 = 0$ or as small as possible.

Example. — Test whether the general equation (7) or (8) is satisfied in the positions shown on Pl. VI.

---

* A design is possible which will cause F_1, for certain web members, to fall within the span but without the panel C_1D_1. This does not occur in any ordinary form of truss.

153. Truss with Parallel Chords. — The application of the results of the foregoing discussion to the case of trusses with parallel chords presents no difficulty.* The only matter calling for special remark relates to the case of web members.

Referring to (B), Pl. VI, it will be seen that if the chords were parallel the point F_1 (and therefore also F and F') would fall at infinity. In order that $C'D'$ and $A'B'$ might intersect in F', these two lines would therefore need to be parallel. The condition for maximum stress in a web member therefore reduces, in case of parallel chords, to the condition that $A'B'$ and $C'D'$ shall be parallel.

Further, the intercept FG and the moment-arm $F_1 J$ become infinite if the chords are parallel, so that the moment-equation takes an indeterminate form. The stress in the member HD_1 can, however, be easily determined by resolving vertically all forces acting upon the portion of the truss to the left of a section cutting the member in question and two chord members. See Art. 142.

The algebraic formula expressing the condition for maximum stress in a web member (equation (8), Art. 152) reduces to simpler form in case of parallel chords. If F_1 passes to infinity, l_1 and n_1 approach a ratio of equality while $\dfrac{n}{n_1}$ approaches zero; so that the equation becomes, in the limit,

$$Q = \frac{n}{l} W.$$

The load P_1 does not enter this equation, but is to be taken as small as possible in applying it.

§ 4. *Truss with Subordinate Bracing.*

154. Description of Truss. — In Fig. 49, Pl. VIII, is represented a form of truss designed to offer points of support for the floor system intermediate between the main panel joints.

---

* For an exhaustive treatment of this subject consult a paper by Professor Henry T. Eddy, Trans. Am. Soc. C. E., Vol. XXII (1890).

The effect of the subordinate vertical members a and diagonal members b requires some explanation. With the arrangement shown in Fig. 49, the main diagonals are tension members and the main verticals compression members, while the subordinate verticals sustain tension and the subordinate diagonals compression. In the form shown in Fig. 50 (Pl. VIII) the subordinate diagonals sustain tension. The members represented by the broken lines in both Fig. 49 and Fig. 50 do not act when the truss sustains dead loads only. Such members will in general be needed only in panels near the middle of the span. In Fig. 49 the two upper diagonals (and in Fig. 50 the two lower diagonals) in each main panel are designed for tension only; the one represented by the broken line comes into action only when the loading is such that the other would be thrown into compression.

155. **Determinateness of Stresses.** — Disregarding counterbraces, it is easily seen that the stresses in all members are determinate.

In the case of members such as e, f, or g (Fig. 49), the determinateness follows from the fact that any such member can be made one of three through which a section may be passed completely dividing the truss.

The stress in g' is determinate, being always equal to that in g. For a like reason, the stress in a is determinate, being always equal to the load carried at C_1.

Considering the point of intersection of a, b, d, e, since the stresses in a and e are determinate, it follows that the stresses in b and d are also determinate; the force polygon for any four forces in equilibrium being completely determinate when two of the forces are known completely, and the directions of the other two are given. This shows that the stresses in b and d would be determinate even if d and e were not collinear.

Like reasoning applied to an upper joint shows that the stress in any main vertical h is determinate.

156. Effect of Subordinate Braces on Stresses in Main Members. — *Upper chord.* — It may be shown that the stresses in the upper chord members have, for any position of the loads, the same values they would have if the subordinate members were not present. Thus, applying the method of moments in the usual manner, the center of moments for the member f* is the point D_1. The stress is determined by dividing the truss by a section through f, e, and g, and applying the conditions of equilibrium to either portion of the truss. For a load between A_1 and F_1, consider the right portion of the truss. The reaction at B_1 is the same, whatever the division into panels; hence the stress is the same as if the members a and b were not present. For a load between B_1 and D_1, the same conclusion is reached by considering the left portion of the truss. For a load between F_1 and D_1 the moment equation for the right portion of the truss will contain the reaction at B_1, and the portion of the load carried at D_1; and this portion is changed by the subdivision of the panel $F_1 D_1$. The moment of the load at D_1 is, however, zero; so that the subdivision of the panel has no effect upon the moment equation nor upon the stress in the member f.

Web members. — At any upper joint of the truss four forces are in equilibrium, — two due to the chord stresses and two to the stresses in the web members. Any two of these being given, the other two are determined. Since the chord stresses have the same value whether the subordinate members are present or not, the same is true of the web stresses. In determining the stresses in d and h, the subordinate members a and b may therefore be disregarded.

The stress in e is changed by the presence of the subordinate members if any load is between F_1 and D_1. The value of the stress, may, however, be determined by the method of sections, the center of moments being at the intersection of f and g. The construction shown at (B), Pl. VI, may be applied to the

---

* The present discussion refers always to **Fig. 49**, Pl. VIII.

member e, the short panel C_1D_1 (Fig. 49) corresponding to the panel C_1D_1 in Pl. VI, and the point F_1 being at the intersection of f and g.

Lower chord. — The analysis just applied to the member e is applicable also to a lower chord member, such as g. Taking a section through f, e, and g, the origin of moments is to be taken at the intersection of e and f. The construction shown on Pl. VI may therefore be applied, the panel marked C_1D_1 in Fig. 49 corresponding to C_1D_1 in Pl. VI.

Since the load at C_1 and the member a are collinear, the stress in g' is always equal to that in g.

157. Condition for Maximum Stress in Main Truss Members. — From the results of Art. 156, it follows that the method already explained for determining what position of the loads causes a maximum stress may be applied to all the main members of the truss shown in Fig. 49. The condition in all cases is that the points F' and G', determined as on Pl. V or Pl. VI, shall coincide. In case of certain members, however, the position of the line $C'D'$ is changed by the presence of the subordinate members. To summarize:

In the case of the members f, d, and h (Fig. 49), the subordinate members are to be disregarded in applying the condition for maximum stress; the whole panel F_1D_1 takes the place of C_1D_1 in the construction on Pl. VI. This principle applies to all upper chord members, all main verticals, and the upper portions of all main diagonals.

In the case of the members e and g, the short panel marked C_1D_1 in Fig. 49 takes the place of C_1D_1 in the construction on Pl. VI; this construction being otherwise unchanged. This applies to all lower chord members, and to the lower portions of the main diagonals.

There is no difficulty in applying similar reasoning to the truss shown in Fig. 50.

158. Subordinate Members. — When the position of the load series is known, the stresses in the subordinate members (a and b, Fig. 49) are easily determined.

The stress in a is always equal to the load carried at C_1.

The members d and e being collinear, the resolved parts of the stresses in a and b perpendicular to d and e must be equal; the stress in b can therefore be found directly from that in a.

It remains to determine the position of the load series which makes the load at C_1 (and therefore the stresses in a and b) a maximum.

Assuming that a load on any panel is carried to the truss wholly at the ends of that panel, the condition for maximum load at any joint may be deduced without difficulty.

In Fig. 51 are represented the load polygon and funicular polygon for a series of loads, A_1F_1 and F_1B_1 being consecutive

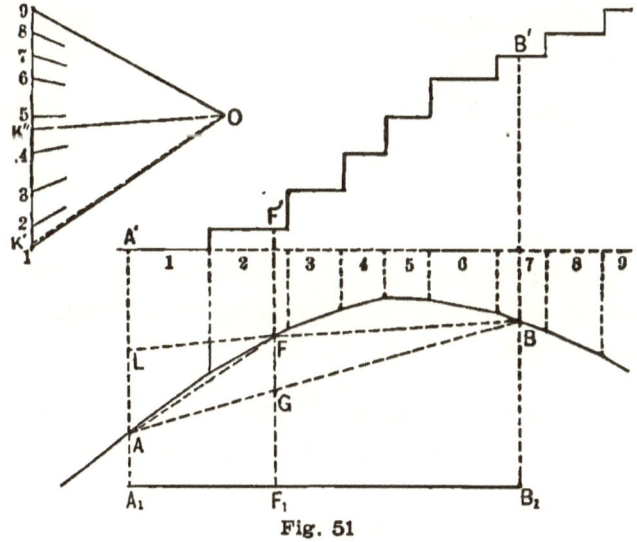

Fig. 51

panels of any lengths. Vertical lines through A_1, B_1, and F_1 intersect the funicular polygon in A, B, F, and the load polygon in A', B', F'. The portion of the loads on A_1F_1 which is carried at F_1 is found by drawing the closing string AF, and parallel to it the ray OK'; the portion carried at F_1 is $K'2$. Similarly, drawing OK'' parallel to the closing string FB, $2\,K''$ represents the portion of the loads on F_1B_1 which is carried at F_1. The total load carried at F_1 is therefore $K'K''$.

Prolonging BF to L, AFL and $K'OK''$ are similar triangles; hence AL and $K'K''$ vary in the same ratio, as the load series moves. But if G is the point in which a vertical through F intersects AB, it is seen that FG varies as AL. Therefore, when $K'K''$ is a maximum, FG is a maximum.

Reasoning as in Arts. 131 and 132, it is found that when FG is a maximum or a minimum the points A', F', and B' are collinear.

The condition for greatest load at F_1 is therefore exactly analogous to the condition for greatest bending moment at F_1 on the supposition that A_1B_1 is a simple beam supported at A_1 and B_1.

From this analogy the conditions for greatest stresses in the members a and b (Fig. 49) may be stated as follows:*

(a) The panels F_1C_1 and C_1D_1 must be loaded as heavily as possible, the heaviest loads being near C_1.

(b) The loads on the panels must be proportional to their lengths.

In satisfying requirement (b), a load must generally be at C_1, and must be regarded as divided between the panels in some arbitrary manner.

159. **Effect of Counterbraces.** — The foregoing discussion and conclusions need modification when the loading is such as to bring counterbraces into action. It is needful to determine what counters are needed, what maximum stresses they sustain, and whether the maximum stresses in any other members are changed by their action. Let it be assumed that when the

---

*See paper by Professor H. T. Eddy, Trans. Am. Soc. C. E., Vol. XXII.

TRUSS WITH SUBORDINATE BRACING.

train moves from right to left the members called into action are those represented in Fig. 52. The conclusions drawn in Art. 155 as to the determinateness of the stresses are equally applicable to the present case; but the fact that the members d and e are not collinear makes it necessary to modify

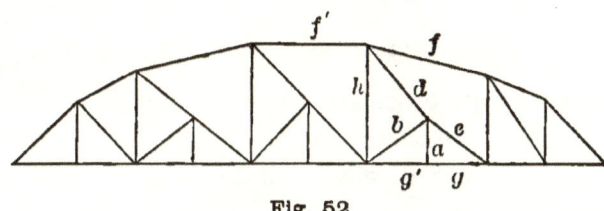

Fig. 52

the methods of determining greatest stresses in certain members.

In case of the members f, g, e, and a (Fig. 52) the usual methods obviously apply.

If b is designed for tension only, it is necessary to determine whether it acts when the stress in d is greatest. If it does not, the discussion of d is obvious. If b does act, the stress in d is not the same as in the case in which d and e are collinear. It may be shown, however, that the stress in d always bears a constant ratio to that which would exist if d and e were collinear. This becomes evident if the force polygon be drawn for the forces due to the stresses in f, f', d, and h, and the effect of changing the direction of d be observed; it being remembered that such change does not affect the stresses in f, f'.

In a truss of long span, the stress in d will usually be small, since the dead load will be so great as to nearly or quite counterbalance the greatest live load effect. Counters will be needed, if at all, only in panels near the middle of the span.

The determination of the maximum stress in b with the counter d in action is less simple than it would be if d and e were collinear; but the member b will have its true maximum when the member d does not act, the train approaching from the left.

160. Influence Lines. — From the foregoing discussion it is seen that the influence lines for the main truss members may be drawn by the general method given in Art. 147. In one case, however, there is a peculiarity which is worthy of special notice.

Considering the lower chord member $C_1 D_1$ (Fig. 49, Pl. VIII), the point F_1 falls between A_1 and C_1, making n_1 negative. The ordinate $F_2 f_2$ is made equal to $\frac{l_1 l_2}{lt}$, positive; $A_2 f_2$ and $f_2 B_2$ are drawn, intersecting verticals through C_2 and D_2 in C_3 and D_3, respectively; the influence line is $A_2 C_3 D_3 B_2$. This peculiar form does not occur in any common form of truss without subordinate bracing.

Figure 50 shows the influence line for the upper chord member g; the subordinate braces being so placed that the diagonals are in tension. Here n_2 is negative, F_1 falling between D_1 and B_1.

For the subordinate members the influence lines may be drawn without difficulty from the principles of Art. 158.

161. Application. — On Pl. VII is shown a truss of 300 ft. span, divided into ten panels of 30 ft. each. The lengths of the main verticals are 30 ft., 45 ft., and 60 ft. The subordinate diagonals are so placed as to sustain tension. Counterbraces are shown in the central panel only; a full computation of stresses due to live and dead loads will determine whether other counters are needed. The moving load is the same as shown on Pl. III.

At (A), Pl. VII, is shown the position of the truss causing greatest stress in the upper chord member $C_1' F_1'$, and at (B) the position causing greatest stress in the web member $C_1' D_1'$. The letters A, B, C, D, F have the same meanings as on Plates V and VI, and the construction is self-explanatory. If the lower part of the web member $C_1' K$, in Fig. (B), were under consideration, the point F_1 would be as shown in the figure, but $C_1 D_1$ would be the double panel $C_1 K$ (the same as if the subordinate members were omitted). These statements are in accordance with the foregoing discussion.

In each of the cases shown on Pl. VII it will be seen that the lines $A'B'$ and $C'D'$ intersect at a point F' lying in the vertical through F_1.

The stresses corresponding to the positions thus determined may be computed by the method of moments, as on Pl. VI. The construction is not shown; but satisfactory results may be secured by careful work and by the use of suitable linear and force scales.

The construction for determining the position for greatest stress in a subordinate member is not shown; it is identical with the construction shown on Pl. IV for determining the position for greatest bending moment at a given section of a beam.

§ 5. *Uniformly Distributed Moving Load.*

162. **Application of General Method.** — The case in which the moving load is assumed to have a uniform horizontal distribution involves no principles additional to those already explained. The foregoing discussion applies to any load distribution whatever. It is, however, of interest to notice the special form assumed by the general principles when the load is uniformly distributed.

The load polygon becomes a straight line whose slope depends upon the intensity of the loading (the load per unit length). The funicular polygon becomes a continuous curve, being in fact a parabola with axis vertical, the position of the vertex depending upon the position of the pole of the force diagram. For the purpose of drawing the funicular polygon the load may be replaced by a series of concentrated loads; the sides of the resulting polygon will be tangent to the desired curve, and the polygon will coincide more and more nearly with the desired parabola, the smaller and nearer together the concentrated loads are taken.

It is obvious that for maximum chord stresses the only condition is that the moving load shall cover the whole span.

For maximum stress in any web member, the head of the load series must be at some point in the corresponding panel; the exact position may be determined by applying the general method. The position of the head of the load series is shown at a glance by the influence diagram. Thus, from diagram (C), Fig. 48 (Pl. VIII), it is obvious that the greatest tension in the member $C_1'D_1$ will occur when the load covers E_1B_1 completely, E_1A_1 being free from load.

163. Assumption of Full Panel Loads. — The determination of the maximum web stresses is somewhat simplified by an approximate assumption as to the way in which the uniformly distributed load comes upon the truss. Thus, in Fig. 53, let it be required to determine the greatest tension in the member BB'. The moving load must be brought on from the right until its head is at some point between A and B. In this position, the loads actually supported at the various joints are as follows:

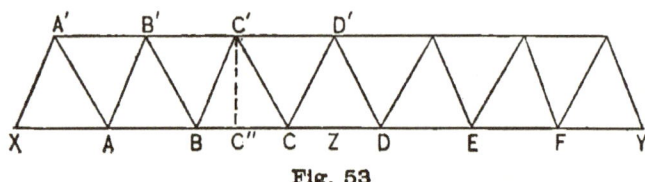

Fig. 53

At C a full panel load (half the load between B and C, and half that between C and D); at each of the points D, E, and F also a full panel load; at B less than a full panel load (half the load on BC, and a portion of that on AB). Let it be assumed that, when the tension in BB' is greatest, all joints from B to F inclusive sustain full panel loads, while all joints to the left of B (only one in the case shown) are free from loads.

If a similar assumption is made in case of every chord member, the resulting stresses will not differ greatly from those obtained by rigorously adhering to the assumption of a uniformly distributed load, the error being on the side of safety. It is to be remarked, also, that the case of a strictly uniform distribution is not realized in practice, so that the results of the above assumption are probably as reliable as would be obtained by following out strictly the assumption of uniform distribution.

The method of determining maximum stresses, on the above assumption as to the loading, will be illustrated by reference to the truss shown in Fig. 54, Pl. VIII. In this truss the panels are of equal length; but the method applies also to the case of unequal panels, the only difference being that the "full panel loads" for different joints are not equal unless the panels are equal.

The principle of counterbracing will be employed here, the diagonals being constructed to sustain tension only.

164. Dead Load Stresses. — The dead load stresses may be determined by means of a stress diagram, as in the roof-truss problems already treated. Two points should be observed in drawing this diagram. (1) If dead loads are taken to act at upper as well as at lower joints, the force polygon for the loads and reactions must show these forces in the same order as that in which their points of application occur in the perimeter of the truss. (2) The diagonal members assumed to act are taken as all sloping in the same direction.

The reason for the first point is the same as already explained in Art. 90; viz. that unless the forces be taken in the order mentioned, the stress diagram cannot be the true reciprocal of the truss diagram and certain lines will have to be duplicated. The reason for assuming the diagonals as all sloping in the same way is the same as in the case of the roof truss with counterbracing (Arts. 111 and 112).

The dead-load stress diagram is not shown, since its construction involves no principle not already fully explained and illustrated.

165. Stresses Due to Moving Load. — (*a*) *Chord members.* — By applying the method of sections it is easily seen that a load at any point produces tension in every lower chord member and compression in every upper chord member. It follows that the greatest stresses in the chord members will occur when the truss is fully loaded. A convenient method of determining the live load stresses is, therefore, to draw a stress diagram, as in case of fixed loads. This diagram is not shown.

(*b*) *Web members.* — In case of a web member, loads in different positions tend to cause opposite kinds of stress. Thus, considering the member $f'n'$ (Fig. 54, Pl. VIII), a tension is caused in it by a load at either of the points *ab*, *bc*, *cd*, *de*, or *ef*, while a compression is caused by a load at *fg*, *gh*, or *hi*. Similarly, loads at *ab*, *bc*, *cd*, *de*, and *ef* all tend to throw compression on the vertical member $f'm'$, while loads on *fg*, *gh*, *hi* have the opposite tendency.

Therefore, to produce the greatest tension upon $f'n'$ (and compression on $f'm'$), the live load must act only at *ef* and all joints to the right; while to cause the greatest compression on $f'n'$ (and tension upon $f'm'$) the live load must act only at *fg*, *gh*, and *hi*. A similar statement will hold regarding any other web member. Since counterbraces are to be used in all panels in which diagonal members would otherwise be thrown into compression, we shall need only to consider the greatest tension in each diagonal and the greatest compression in each vertical. We shall first outline the method to be employed, and then explain the construction.

To determine the greatest tension in a diagonal member, as $g'm'$: Assume the live load to come upon the bridge from the right until there are full loads at the joints *ab*, *bc*, *cd*, *de*, *ef*, and *fg*. Take a section cutting $g'm'$ and the two chord members gg' and $m'm$, and consider the forces acting upon the portion of the truss to the left of the section. These forces are four in number: the reaction at the support and the forces acting in the three members cut. Hence we first determine the reaction, and then determine the three other forces for equilibrium by the method of Art. 42.

The construction is shown in Fig. 54 (Pl. VIII). *ABCDEFGHI* is the force polygon for the eight live loads that may come upon the truss. Choosing a pole O, the funicular polygon for the eight loads is next drawn. Now, turning the attention to the member $g'm'$, the loads gh and hi are assumed not to act. The reactions at the supports for this case of loading are found in the usual way. Prolong oa and og to intersect the lines of action of the two reactions, and join the two points thus determined. This gives the closing line of the funicular polygon (or om). The ray OM is now drawn parallel to the string om, and the two reactions are GM and MA.

Now take a section through mm', $m'g'$, and $g'g$, and apply the construction of Art. 42 to the determination of the forces acting in the three members cut. The resultant of GM and the force in gg' must act through the point X (the intersection of gg' produced and ij). The resultant of the forces in mm' and $m'g'$ must act through their intersection Y. Hence these two resultants (being in equilibrium with each other) must both act in the line XY. From G draw a line parallel to gg', and from M a line parallel to XY; mark their point of intersection G'. Then MG' is the resultant of the forces in mm' and $m'g'$. From M draw a line parallel to mm', and from G' a line parallel to the member $g'm'$, and mark their point of intersection M'. Then $M'G'$ represents the force in the member $m'g'$. This is also the value of the greatest stress in $m'g'$.

To determine the stress in the vertical member $l'g'$, the same loading must be assumed, and a similar construction is employed. Take a section cutting ll', $l'g'$, and $g'g$, and determine forces acting in these three lines which shall be in equilibrium with the reaction GM. This reaction is in equilibrium with MG' and $G'G$, the former having the line of action XY. Then MG' is resolved into two forces having the directions of the members $g'l'$ and $l'l$. The stress in $g'l'$ is found to be a compression, represented in the stress diagram by the line $G'L'$.

The maximum live load stress in every web member may be found in the same way. If the above reasoning is understood, there will be no difficulty in applying the same method to the remaining members.

166. **Maximum Stresses.** — By combining the stresses due to live and dead loads, the maximum stresses are easily determined.

Web members. — When the web members are considered, the effect of counterbracing needs careful attention.

The construction above explained gives the greatest live load tension in each diagonal. It may be that for certain members, this tension is wholly counterbalanced by the dead loads. In any panel in which this is the case, the member shown will never act and may be omitted. The counterbrace must then be considered.

Evidently, the algebraic sum of the stresses due to live and dead loads will be the true maximum tension in each of the diagonals shown. For the greatest tension in the other system of diagonals, the load must be brought on the bridge from the left; or, what amounts to the same thing, the tension already found for any one of the diagonals shown is also the greatest tension in the diagonal sloping the opposite way in the panel equally distant from the middle of the truss. (In fact, the stresses in all members due to a movement of the loads from left to right may be obtained from the results already reached, by consideration of symmetry.)

As to the vertical members, two values of the stress must be compared in every case — namely, the greatest compressions corresponding to the two directions of the moving load. But both can be obtained from the above results by considering the symmetry of the truss. For example, the stress found for $l'g'$ is a possible value for the stress in $r'b'$, and must be compared

UNIFORMLY DISTRIBUTED MOVING LOAD.

with the value obtained for the latter member when the load moves from right to left. It is possible, also, that certain of the verticals may be in tension when the dead loads act alone.

Maximum chord stresses. — These are found by combining the stresses due to fixed and moving loads, determined as already described.

PART III.

CENTROIDS AND MOMENTS OF INERTIA.

CHAPTER VIII. CENTROIDS.

§ 1. *Centroid of Parallel Forces.*

167. Composition of Parallel Forces. — The composition of complanar parallel forces can always be effected by means of the funicular polygon, by the method of Art. 27. It is now necessary to consider parallel systems more at length, as a preliminary to the discussion of graphic methods for determining centers of gravity and moments of inertia.

168. Resultant of Two Parallel Forces. — Let ab and bc (Fig. 55) be the lines of action of two parallel forces, their magnitudes being AB and BC (not shown). Let ac be the line of action of their resultant, and AC (not shown) its magnitude. By the principle of moments (Art. 50) the sum of the moments of AB and BC about any point in their plane is equal to the moment of AC about the same point. If the origin of moments is on ac, the moment of AC is zero; and therefore the moments of AB and BC are numerically equal (but of opposite signs).

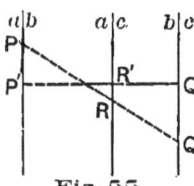

Fig. 55

Let any line be drawn perpendicular to the given forces, intersecting their lines of action in P', Q', and R' respectively. Then $AB \times P'R' = BC \times Q'R'$.

CENTROID OF PARALLEL FORCES.

Let any other line be drawn intersecting the three lines of action in P, Q, and R respectively. Then

$$\frac{PR}{P'R'} = \frac{QR}{Q'R'},$$

and therefore $\qquad AB \times PR = BC \times QR.$

That is, PQ is divided by the line ac into segments inversely proportional to AB and BC.

If AB and BC act in opposite directions, the line ac will be outside the space included between ab and bc; but the above result is true for either case.

169. Centroid of Two Parallel Forces. — If the lines of action ab and bc (Fig. 55) be turned through any angle about the points P and Q respectively, the forces remaining parallel and of unchanged magnitudes, the line of action of their resultant will always pass through the point R. For, by the preceding article, the line of action of the resultant will always intersect PQ in a point which divides PQ into segments inversely proportional to AB and BC. Hence, if AB and BC remain unchanged, and also the points P and Q, the point R must also remain fixed.

If P and Q are taken as the points of application of AB and BC, R may be taken as the point of application of AC, in whatever direction the parallel forces are supposed to act. The point R is called the centroid * of the parallel forces AB and BC for the fixed points of application P and Q.

170. Centroid of Any Number of Parallel Forces. — Let AB, BC, and CD be three parallel forces, and let P, Q, and S (Fig. 56) be their fixed points of application. Let R be the centroid of AB and BC, and AC their resultant. Take R as the fixed point of application of AC, and determine T, the centroid of AC

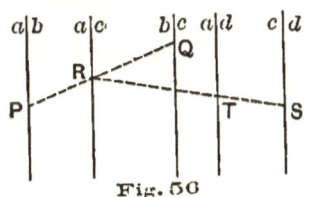

Fig. 56

---

* The name *center of parallel forces* has been quite commonly used instead of *centroid* as above defined. The latter term has, however, been adopted by some of the later writers, and seems on the whole a better designation.

and *CD*. Let *AD* be the resultant of *AC* and *CD*; then *AD* is also the resultant of *AB*, *BC*, and *CD*.

Now if *AB*, *BC*, and *CD* have their direction changed, but still remain parallel and unchanged in magnitude, it is evident that the point *T*, determined as above, will remain fixed and will always be on *ad*, the line of action of *AD*. The point *T* is called the centroid of the three forces *AB*, *BC*, and *CD*.

By an extension of the same method, a *centroid* may be determined for any system of parallel forces having fixed points of application. Hence the following definition may be stated:

The centroid of a system of parallel forces having fixed points of application is a point through which the line of action of their resultant passes, in whatever direction the forces be taken to act.

In determining the centroid by the method just described, the forces may be taken in any order without changing the result. For the centroid must lie on the line of action of the resultant; and since this is a determinate line for each direction in which the forces may be taken to act, there can be but one centroid.

171. Non-complanar Parallel Forces. — The reasoning of the preceding articles is equally true, whether the forces are complanar or not. In what follows we shall deal either with complanar forces, or with forces whose *points of application are complanar*. No more general case will be discussed.

172. Graphic Determination of Centroid of Parallel Forces. — If the line of action of the resultant of any system of parallel forces be found for each of two assumed directions of the forces, the point of intersection of these two lines is the centroid of the system. Moreover, if the points of application are complanar, the two assumed directions may both be such that the forces will be complanar.

Thus, let the plane of the paper be the plane containing the given points of application, and let *ab*, *bc*, *cd*, *dc*, *cf* (Fig. 57) be the points of application of five parallel forces, *AB, BC, CD, DE, EF*. Draw through these points parallel lines in some chosen direction, and taking them as the lines of action of the given forces, construct the force and funicular polygons corre-

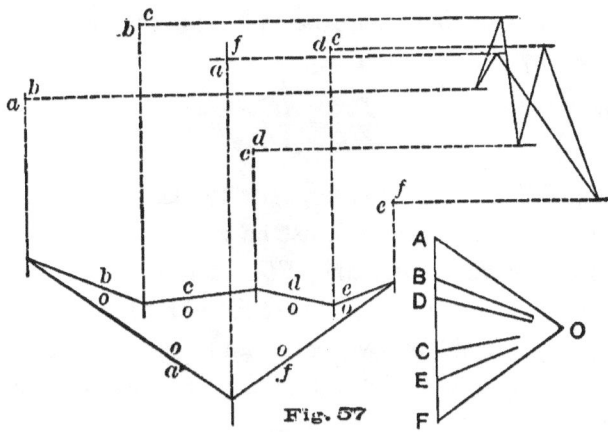

Fig. 57

sponding. The line of action of the resultant is drawn through the intersection of the strings *oa* and *of*, and this line contains the centroid of the given forces. Next draw through the given points of application another set of parallel lines, preferably perpendicular to the set first drawn, and draw a funicular polygon for the given forces with these lines of action. (It is unnecessary to draw a new force diagram, since the strings in the second funicular polygon may be drawn respectively perpendicular to those of the first.) This construction determines a second line as the line of action of the resultant corresponding to the second direction of the forces. The required centroid of the system is the point of intersection of the two lines of action of the resultant thus determined, and is the point *af* in the figure.

Example. — Find graphically the centroid of the following system of parallel forces, and test the result by algebraic computation: A force of 20 lbs. applied at a point whose rectangular coördinates are (4, 6); 12 lbs. at the point (12, 3); 20 lbs. at the point (10, 10); −10 lbs. at the point (7, −9); −8 lbs. at the point (−5, −10).

173. Centroid of a Couple.—If AB and BC are the magnitudes of any two parallel forces applied at points P and Q, then by Art. 169 their centroid R is on the line PQ and is determined by the equation

$$\frac{PR}{QR} = \frac{BC}{AB}.$$

If AB and BC are equal and opposite forces, PR and QR must be numerically equal, and R must be *outside* the space between the lines of action of AB and BC. These conditions can be satisfied only by making PR and QR infinite. Hence we may say that the centroid of two equal and opposite forces lies on the line joining their points of application and is infinitely distant from these points.

174. Centroid of a System whose Resultant is a Couple.— If a system with fixed points of application is equivalent to a couple, its centroid will be infinitely distant from the given points of application. A line containing this centroid can be determined as follows:

Take the given forces in two groups; the resultants of the two groups will be equal and opposite. Find the centroid of each group, and suppose each partial resultant applied at the corresponding centroid. Then the centroid of the whole system is the same as that of the couple thus formed, and will lie on the line joining the two partial centroids.

If the separation into groups be made in different ways, different couples and different partial centroids will be found. The different couples are, of course, equivalent; and it may be proved that the lines joining the different pairs of partial centroids are all parallel, and intersect in the (infinitely distant) centroid of the whole system.

For, suppose the given system to be equivalent to a couple

Q with points of application A and B, and also to a couple Q' with points of application A' and B'. These two couples must be equivalent to each other, whatever be the direction of the forces. Let AB be taken as this direction. The two equal and opposite forces of the couple Q neutralize each other, since their lines of action are coincident; hence the two forces of the couple Q' must also neutralize each other. Therefore $A'B'$ must be parallel to AB.

175. **Moment of a Force about an Axis.** — The moment of a force with respect to a given axis, as defined in Art. 47, depends not only upon the point of application of the force, but upon its direction. In dealing with systems of forces whose direction may change but whose points of application are complanar, we shall need to compute moments only for axes lying in the plane of the points of application; and the forces may usually be regarded as acting in lines perpendicular to this plane. Hence we shall compute moments by the following rule:

The moment of a force with reference to any axis is the product of the magnitude of the force into the distance of its point of application from the axis.

§ 2. *Center of Gravity — Definitions and General Principles.*

176. **Center of Gravity of Any Body.** — Every particle of a terrestrial body is attracted by the earth with a force proportional directly to the mass of the particle. The resultant of such forces upon all the particles of a body is its *weight;* and the point of application of this resultant is called the *center of gravity* of the body. The lines of action of these forces may be assumed parallel without appreciable error. We may therefore define the center of gravity of a body as follows:

If forces be supposed to act in the same direction upon all particles of a body, each force being proportional to the mass of the particle upon which it acts, the centroid of this system of parallel forces is the center of gravity of the body.

This point is also called *center of mass*, and *center of inertia*, either of which is a better designation than center of gravity. The latter term is, however, in more general use.

177. Centers of Gravity of Areas and Lines. — The term center of gravity, as above defined, has no meaning when applied to lines and areas, since these magnitudes have no mass, and hence are not acted upon by the force of gravity. It is, however, common to use the name center of gravity in the case of lines and areas, with meanings which may be stated as follows:

The *center of gravity of an area* is the center of gravity of its mass, on the supposition that each superficial element has a mass proportional to its area. This point would be better described as the *center of area*.

The *center of gravity of a line* is the center of gravity of its mass, on the assumption that each linear element has a mass proportional to its length. The term *center of length* is preferable, and will often be used in what follows.

Similar statements might be made regarding geometrical solids, but we shall have to deal chiefly with lines and areas.

178. Moments of Areas and Lines. — *Definition*. — The moment of a plane area with reference to an axis lying in its plane is the product of the area by the distance of its center of gravity from the axis.

Proposition. — The moment of any area about a given axis is equal to the sum of the moments of any set of partial areas into which it may be divided. For, by the definition of center of gravity, a force numerically equal to the total area and applied at its center of gravity is the resultant of a system of forces numerically equal to the partial areas and applied at their respective centers of gravity; and the moment of any force is equal to the sum of the moments of its components (Art. 50).

A similar definition and proposition may be stated regarding lines.

The moment of an area or line is zero for any axis containing its center of gravity.

179. Symmetry. — Two points are symmetrically situated with respect to a third point if the line joining them is bisected by that point.

Two points are symmetrically situated with respect to a line or plane when the line joining them is perpendicular to the given line or plane and bisected by it.

A body is symmetrical with respect to a point, a line, or a plane, if for every point in the body there is another such that the two are symmetrically situated with respect to the given point, line, or plane. The point, line, or plane is in this case called a point of symmetry, an axis of symmetry, or a plane of symmetry of the body.

180. General Principles. — (1) The center of gravity of two masses taken together is on the line joining the centers of gravity of the separate masses. For it is the point of application of the resultant of two parallel forces applied at those points.

(2) If a body of uniform density has a plane of symmetry, the center of gravity lies in this plane. If there is an axis of symmetry, the center of gravity lies in this axis. If the body is symmetrical with respect to a point, that point is the center of gravity. For the elementary portions of the body may be taken in pairs such that for each pair the center of gravity is in the plane, axis, or point of symmetry.

181. Centroid. — The center of gravity of any body or geometrical magnitude is by definition the same as the centroid of a certain system of parallel forces. It will be convenient, therefore, to use the word centroid in most cases instead of center of gravity.

§ 3. *Centroids of Lines and of Areas.*

182. General Method of Finding Centroid. — The centroid of any area may be found by the following method: Divide the given area into parts such that the area and centroid of each part are known. Take the centroids of the partial areas as the points of application of forces proportional respectively to those areas. The centroid of this system of forces is the centroid of the total area, and may be found by the method of Art. 172.

The centroid of a line may be found by a similar method.

The method just described is exact if the magnitudes of the partial areas and their centroids are accurately known. If the given area is such that it cannot be divided into known parts, it will still be possible to get an approximate result by this method.

In applying this general method, it is frequently necessary to know the centroids of certain geometrical lines and figures, and also the relative magnitudes of the areas of such figures. Methods of determining these will be given in the following articles.

183. Centroids of Lines. — The centroid of a *straight line* is at its middle point.

Broken line. — The centroid of a broken line is the center of a system of parallel forces, of magnitudes proportional to the lengths of the straight portions of the line, and applied respectively at their middle points. It may be found graphically by the method of Art. 172, or by any other method applicable to parallel forces.

Part of regular polygon. — For the centroid of a part of a regular polygon, a special construction is found useful.

CENTROIDS OF LINES AND OF AREAS.

Let $ABCDE$ (Fig. 58) be part of such a polygon, and O the center of the inscribed circle. Let

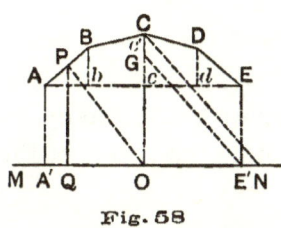

Fig. 58

r = radius of inscribed circle; l = length of a side of the polygon; s = total length of the broken line AE. Through O draw OC, the axis of symmetry of AE; and MN, perpendicular to OC.

First, the centroid must lie on OC.

Second, to find its distance from O, assume a system of equal and parallel forces applied at the middle points of the sides AB, BC, etc. The required centroid is the point of application of the resultant of these forces. Taking MN as the axis of moments, and letting x = required distance of centroid from MN, and x_1, x_2, x_3, x_4 the distances of the middle points of AB, BC, etc., from MN, we have from the principle of moments,

$$lx_1 + lx_2 + lx_3 + lx_4 = sx.$$

But if Ab and Bb be drawn perpendicular respectively to PQ and MN we have from the similar triangles ABb, POQ,

$$\frac{AB}{Ab} = \frac{PO}{PQ} \text{ or } \frac{l}{Ab} = \frac{r}{x_1}.$$

$$\therefore lx_1 = r \cdot Ab.$$

In the same way,

$$lx_2 = r \cdot bc,$$

$$lx_3 = r \cdot cd,$$

$$lx_4 = r \cdot dE.$$

Hence, $s \cdot x = r (Ab + bc + cd + dE) = r \cdot AE$,

where AE is equal to the projection of the broken line $ABCDE$ on MN.

The centroid G may now be found graphically as follows: Make $Oc' = r$; $ON = \frac{1}{2} s$; $OE' = \frac{1}{2} AE$; draw Nc'. Then G is determined by drawing $E'G$ parallel to Nc'.

Circular arc. — The above construction holds, whatever the length of the side *l*. If this length be decreased indefinitely, while the number of sides is increased indefinitely, so that the length *s* remains finite, we reach as the limiting case a circular arc. The same construction therefore applies to the determination of the centroid of such an arc, *r* denoting the radius of the circle and *s* the length of the arc.

184. **Centroids of Geometrical Areas.** — *Parallelogram.* — The centroid of a parallelogram is on a line bisecting two opposite sides.

Let *ABCD* (Fig. 59) be a parallelogram, and *EF* a line bisecting *AD* and *BC*. Divide *AB* into any even number of equal parts, and through the points of division draw lines parallel to *BC*. Also divide *BC* into any even number of equal parts and draw through the points of division lines parallel to *AB*. The given parallelogram is thus divided into equal elements. Now

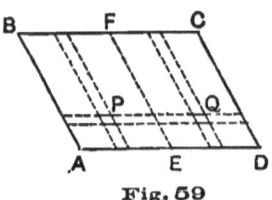

Fig. 59

consider a pair of these elements, such as those marked *P* and *Q* in the figure, equally distant from *AD*, and also equally distant from *EF*, but on opposite sides of it. The centroid of the two elements taken together is at the middle point of the line joining their separate centroids. If the number of divisions of *AB* and of *BC* be increased without limit, the elements approach zero in area, and the centroids of *P* and *Q* evidently approach two points which are equally distant from *EF*. Hence in the limit, the centroid of such a pair of elements lies on the line *EF*. But the whole area *ABCD* is made up of such pairs; hence the centroid of the whole area is on the line *EF*. For like reasons it is also on the line bisecting *AB* and *DC*; hence it is at the intersection of the two bisectors.

The point thus determined evidently coincides with the point of intersection of the diagonals *AC* and *BD*.

Triangle. — The centroid of a triangle lies on a line drawn from any vertex to the middle of the opposite side; and is, therefore, the point of intersection of the three such lines.

Let ABC (Fig. 60) be any triangle, and D the middle point of BC. Then the centroid of ABC must lie on AD. For AD bisects all lines, such as bc, parallel to BC. Now inscribe in the triangle any number of parallelograms such as $bcc'b'$, with sides parallel respectively to BC and AD. The centroid of each parallelogram lies on AD, and, therefore, so also does the centroid of the whole area composed of such parallelograms. If the number of such parallelograms be increased without limit, the alti-

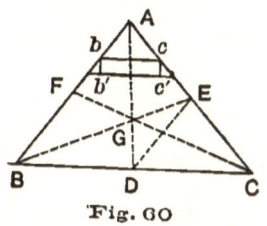

Fig. 60

tude of each being diminished without limit, their combined area will approach that of the triangle, and the centroid of this area will approach in position that of the triangle. But since the former point is always on the line AD, its limiting position must be on that line. Therefore the line AD contains the centroid of the triangle.

By the same reasoning, it follows that the centroid of ABC must lie on BE, drawn from B to the middle point of AC. Hence it must be the point of intersection of AD and BE, which point must also lie on the line CF drawn from C to the middle point of AB.

The point G divides each bisector into segments which are to each other as 1 to 2. For, from the similar triangles ABC, EDC, since EC is half of AC, it follows that DE is equal to half of BA. And from the similar triangles AGB, DGE, since DE is half of AB, it follows that GE is half of GB, and GD half of GA.

Quadrilateral. — Let $ABCD$ (Fig. 61) be a quadrilateral of which it is required to find the center of gravity. Draw BD, and let E be its middle point. Make $EG_1 = \frac{1}{3} EA$, and $EG_2 = \frac{1}{3} EC$. Then the centroids of the triangles ABD and BCD are

G_1 and G_2 respectively. Hence the centroid of $ABCD$ is on the line G_1G_2 at a point dividing it into segments inversely proportional to the areas of ABD and BCD. Since these two triangles have a common base, their areas are proportional to their altitudes measured from this base. But these altitudes are proportional to AF and FC, or to G_1H and G_2H; hence, if G is the required centroid,

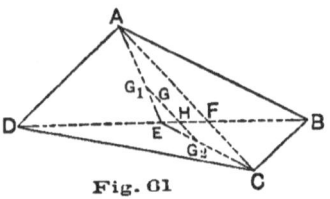

Fig. 61

$$G_1G : G_2G :: G_2H : G_1H.$$

Therefore G is found by making $G_1G = G_2H$.

Circular sector. — To find the centroid of a circular sector OAB (Fig. 62) we may reason as follows: Draw two radii OM, ON, very near together. Then OMN differs little from a triangle, and its centroid will fall very near the arc $A'B'$, drawn with radius equal to two-thirds of OA. If the whole sector be subdivided into elements such as OMN, their centers of gravity will all fall very near to the arc $A'B'$. If the number of such elements is indefinitely increased, the line joining their centroids approaches as a limit the arc $A'B'$. And since the areas of the elements are proportional to the lengths of the corresponding portions of $A'B'$, the centroid of the total area is the same as that of the arc $A'B'$.

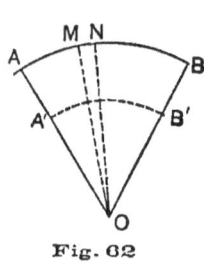

Fig. 62

This point may be found by the method described in Art. 183.

185. **Graphic Determination of Areas.** — Let there be given any number of geometrical figures, and let it be required to determine the relative magnitudes of their areas.

If a number of rectangles can be found, of areas equal respectively to the areas of the given figures and having one common side, then the remaining sides will be proportional to the areas of the given figures.

CENTROIDS OF LINES AND OF AREAS.   165

An important case is that in which the given figures are such that the area of each is equal to the product of two known lines. In this case a series of equivalent rectangles having one common side can be found by the following construction.

Let ABC (Fig. 63) be a triangle, and let it be required to determine an equivalent rectangle having a side of given length as LM. Let b and h be the base and altitude of the triangle. Make $LN = \frac{1}{2}b$ and $LP = h$, and draw the semicircumference PRN. Draw the ordinate LR perpendicular to PN; then

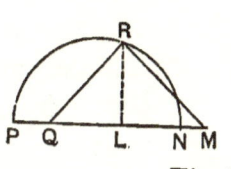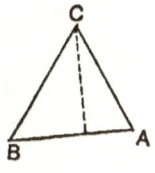

Fig. 63

$$\overline{LR}^2 = PL \times LN = \tfrac{1}{2}\,bh = \text{area } ABC.$$

Draw MR, and perpendicular to it draw RQ. Then

$$\overline{LR}^2 = LM \times LQ = \text{area } ABC.$$

Hence LQ is the required length.

If the given figure is a parallelogram, LN and LP may be its base and altitude. If it be a circular sector, LN and LP may be the length of the arc and half the radius.

186. **Centroids of Partial Areas.** — It may be required to find the centroid of the part remaining after deducting known parts from a given area. For this case the construction of Art. 182 needs modification. In the case there considered we regarded the partial areas and the total area as proportional to parallel forces acting at their respective centers of gravity. In this case we may also represent the total area, the portions deducted from it, and the remaining portion as forces acting at the respective centers of gravity; but the forces corresponding to the areas deducted must be taken as acting in the opposite direction to that assumed for the forces representing the total area and the area remaining.

Thus, to find the centroid of the area remaining after deducting from the circle *ABD* (Fig. 64) a smaller circle *EFH* and a sector *OAB*, we may proceed as follows: Find the centroid of three parallel forces proportional to the areas *ABD*, *EFH*, and *OAB*, and applied at their respective centroids *O*, *C*, and *G*; but the last two must be taken as acting in the direction opposite to that of the first. With this understanding, the force and funicular polygons may be employed as in Art. 172.

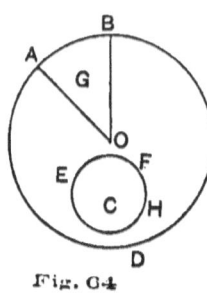
Fig. 64.

187. **Moments of Areas.** — The moment of an area about any line in its plane may be determined from the funicular polygon employed in finding its center of gravity. Let the parallel forces applied at the centroids of the partial areas be assumed to act parallel to the axis of moments. Then the distance intercepted on the axis by the extreme lines of the funicular polygon, multiplied by the pole distance, is equal to the moment of the total area about the axis. For, by Art. 56, this construction gives the moment of the resultant force about any point in the given axis; and this is equal to the moment of the resultant area about the axis by definition.

A similar rule gives the moments of the partial areas.

CHAPTER IX. MOMENTS OF INERTIA.

§ 1. *Moments of Inertia of Forces.*

188. Definitions. — The *moment of inertia of a body* with respect to an axis is the sum of the products obtained by multiplying the mass of every elementary portion of the body by the square of its distance from the given axis.

The *moment of inertia of an area* with respect to an axis is the sum of the products obtained by multiplying each elementary area by the square of its distance from the axis.

The *moment of inertia of a line* may be similarly defined, using elements of length instead of elements of mass or area.

The *moment of inertia of a force* with respect to any axis is the product of the magnitude of the force into the square of the distance of its point of application from the axis. The sum of such products for any system of parallel forces is the moment of inertia of the system with respect to the given axis.

[NOTE. — The term moment of inertia had reference originally to material bodies, the quantity thus designated having especial significance in dynamical problems relating to the rotation of rigid bodies. The quantity above defined as the moment of inertia of an area is of frequent occurrence in the discussion of beams, columns, and shafts in the mechanics of materials. In the graphic discussion of moments of inertia of areas, it is convenient to treat areas as forces, just as in the determination of centers of gravity; it is therefore convenient to use the term *moment of inertia of a force* in the sense above defined. It is only in the case of masses that the term moment of inertia is really appropriate, but it is by analogy convenient to apply it to the other cases.]

The *product of inertia of a mass* with respect to two planes is the sum of the products obtained by multiplying each elementary mass by the product of its distances from the two planes.

The product of inertia of an area, a line, or a force may be defined in a similar manner.

For a *plane area*, the product of inertia with respect to two lines in its plane may be defined as the sum of the products obtained by multiplying each element of area by the product of its distances from the given lines. This is equivalent to the product of inertia of the area with respect to two planes perpendicular to the area and containing the two given lines.

For a *system of forces* with points of application in the same plane, the product of inertia with reference to two axes in that plane may be defined as the sum of the products obtained by multiplying the magnitude of each force by the product of the distances of its point of application from the given axes. It is with such systems and with plane areas that the following pages chiefly deal.

The *radius of gyration* of a body with respect to an axis is the distance from the axis of a point at which, if the whole mass of the body were concentrated, its moment of inertia would be unchanged. The square of the radius of gyration is equal to the quotient obtained by dividing the moment of inertia of the body by its mass.

The radius of gyration of an area may be defined in a similar manner.

The radius of gyration of a system of parallel forces is the distance from the axis of the point at which a force equal in magnitude to their resultant must act in order that its moment of inertia may be the same as that of the system. The square of the radius of gyration may be found by dividing the moment of inertia of the system by the resultant of the forces.

189. Algebraic Expressions for Moment and Product of Inertia. — *Moment of inertia.* — Let m_1, m_2, etc., represent elementary masses of a body, and r_1, r_2, etc., their respective

MOMENTS OF INERTIA OF FORCES.

distances from an axis; then the moment of inertia of the body with respect to that axis is

$$m_1 r_1^2 + m_2 r_2^2 + \cdots \equiv \Sigma m r^2,$$

the symbol Σ being a sign of summation, and the second member of the equation being merely an abbreviated expression for the first.

Product of inertia. — Let p_1, p_2, etc., denote the perpendicular distances of elements m_1, m_2, etc., from one plane, and q_1, q_2, etc., their distances from another; then the product of inertia of the body with respect to the two planes is

$$m_1 p_1 q_1 + m_2 p_2 q_2 + \cdots \equiv \Sigma m p q.$$

Radius of gyration. — With the same notation, if k denotes the radius of gyration of the body, we have

$$(m_1 + m_2 + \cdots) k^2 = m_1 r_1^2 + m_2 r_2^2 + \cdots.$$

Or,

$$k^2 = \frac{m_1 r_1^2 + m_2 r_2^2 + \cdots}{m_1 + m_2 + \cdots} = \frac{\Sigma m r^2}{\Sigma m}.$$

Here Σm is equal to the whole mass of the body.

Product-radius. — Let c represent a quantity defined by the equation

$$(m_1 + m_2 + \cdots) c^2 = m_1 p_1 q_1 + m_2 p_2 q_2 + \cdots.$$

Then if the whole mass were concentrated at the same distance c from two axes, its product of inertia with respect to those axes would be unchanged. This quantity c is thus seen to be analogous to the radius of gyration. It may be called the *product-radius* of the body with respect to the two axes. The value of c is always given by the equation

$$c^2 = \frac{\Sigma m p q}{\Sigma m}.$$

Expressions similar to those just given apply also to plane areas and to systems of parallel forces. In case of an area, m_1, m_2, etc., denote elements of area; r_1, r_2, etc., their distances from the axis of inertia; and (p_1, q_1), (p_2, q_2), etc., their distances from the two planes. In case of a system of parallel forces with complanar points of application, m_1, m_2, etc., must be replaced by the magnitudes of the forces.

190. Determination of Moment of Inertia of a System of Parallel Forces. — Let the points of application of the forces be in the same plane, which also contains the assumed axis. We shall have to deal only with systems satisfying these conditions. By the definition, the moment of inertia will be the same, whatever the direction of the forces. If we take the moment of any force (as defined in Art. 175) about the given axis, and suppose a force equal in magnitude to this moment to act at the point of application of the original force, and in a direction corresponding to the sign of the moment, then the moment of this new force about the given axis is equal to the moment of inertia of the original force. If this be done for all the forces, the algebraic sum of the results will be the required moment of inertia of the system.

This process can be carried out graphically by methods already described.

Let ab, bc, cd, de (Fig. 65) be the points of application of four parallel forces, and let the axis of inertia be QR. Suppose all the forces to act in lines parallel to QR, passing through the given points of application. Their respective moments with reference to QR may now be found by the method of Art. 55.

Draw the force polygon $ABCDE$ and choose a pole O, taking the pole distance H preferably equal to AE or some simple multiple of AE. (In Fig. 65 H is taken equal to AE.) Draw a funicular polygon and prolong each string to intersect QR. Then the moment of any force with respect to QR is the product of H by the distance intercepted on QR by the two strings corresponding to the force in question. Thus the moment of AB is given by the distance $A'B'$ (Fig. 65) multiplied by H. Also the successive moments of BC, CD, DE are represented by $B'C'$, $C'D'$, $D'E'$, each multiplied by H. It is

MOMENTS OF INERTIA OF FORCES. 171

seen also that the intercepts, if read in the above order, give a distinction between positive and negative moments; upward distances on QR denoting in this case positive moments, and downward distances negative moments.

We have now to find the sum of the moments of a second system of forces acting in the original lines, but represented in

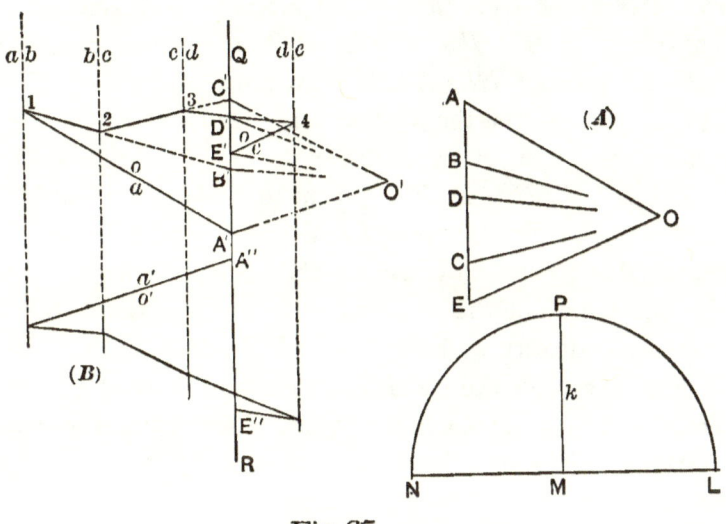

Fig. 65.

magnitude and direction by the intercepts just found. We may take as the force polygon for the second construction the line $A'B'C'D'E'$ (Fig. 65), and choosing any pole distance H', draw a second funicular polygon and find the distances intercepted by the successive strings on the line QR. But in this case, since only the resultant is desired, we need only find the intercept $A''E''$ between the first and last strings. The product of this intercept by H' gives the sum of the moments of $A'B'$, $B'C'$, $C'D'$, $D'E'$ with respect to QR; and, if the product be multiplied by H (since $A'B'$, $B'C'$, etc., should each be multiplied by H in order to represent the magnitudes of the forces of the second system), the result will be the required moment of inertia of the given system of forces AB, BC, CD, DE.

It should be noticed that in Fig. 65 (A) is a *force diagram*, (B) a *space diagram* (Art. 11); that is, every line in (A) represents a force, while every line in (B) represents a distance. Even $A'B'$, $B'C'$, etc., though used as forces, are actually merely distances; and the moment of any one of them is the product of a *length* by a *length*.

191. **Radius of Gyration.** — The moment of inertia of the given system is $H \times H' \times A''E''$. If H has been taken equal to AE, the product $H' \times A''E''$ must equal the square of the radius of gyration of the system with respect to QR. The length of the radius of gyration can be found as follows (Fig. 65): Draw $LN = H' + A''E''$, and make $LM = H'$. On LN as a diameter draw a semicircle LPN, and from M draw a line perpendicular to LN, intersecting the semicircle in P. Then MP is the length of the required radius of gyration. For by elementary geometry we have $\overline{PM}^2 = LM \times MN$.

If H is taken equal to $n \times AE$, the moment of inertia is equal to $H' \times A''E'' \times n \times AE$, and the square of the radius of gyration is equal to $nH' \times A''E''$. Hence in Fig. 65 we should put either $LM = nH'$, or $MN = n \times A''E''$.

192. **Central Axis.** — If the axis with reference to which the moment of inertia is found contains the centroid of the given system of forces, it is called a *central axis* of the system.

In many cases it is desired to find the moment of inertia with respect to a central axis whose direction is known while the position of the centroid is at first unknown. It is to be noticed that the method shown in Fig. 65 is applicable in this case; for the first part of the process is identical with that employed in finding the centroid of the system. If, in Fig. 65, the strings *oa*, *oc*, of the first funicular polygon be prolonged to intersect, a line through their point of intersection, parallel to the direction of the forces, will contain the centroid of the system. If this line is taken as the inertia-axis, the points A', B', C', D', E' are the points in which this axis is intersected by the strings oa, ob, oc, od, oe. No further modification of the process is necessary.

193. Moment of Inertia Determined from Area of Funicular Polygon. — In Fig. 65, the moments of the given forces are represented by the intercepts $A'B'$, $B'C'$, $C'D'$, $D'E'$, each multiplied by the pole distance H. The moment of inertia of any force, as AB, is equal to the moment of a force represented by the corresponding intercept as $A'B'$, supposed to act in the line ab. Now, by definition (Art. 175) the moment about the axis QR of a force equal to $A'B'$ acting in the line ab is equal to double the area of the triangle $A'\,1\,B'$; hence the moment of the force $H \times A'B'$ is equal to double the area of that triangle multiplied by H. Similarly, the moment of a force $H \times B'C'$, acting in the line bc, is equal to double the area of the triangle $B'\,2\,C'$ multiplied by H. Applying the same reasoning to each force, we see that the sum of the moments of the assumed forces ($H \times A'B'$, $H \times B'C'$, etc.) is equal to $2H$ times the sum of the areas of the triangles $A'\,1\,B'$, $B'\,2\,C'$, $C'\,3\,D'$, $D'\,4\,E'$. In adding these triangles, each must be taken with its proper sign, corresponding to the sign of the moment represented by it. Thus, the moments of $A'B'$, $B'C'$, and $D'E'$ all have the same sign, while the moment of $C'D'$ has the opposite sign. The algebraic sum of the areas is, therefore,

area $A'\,1\,B'$ + area $B'\,2\,C'$ − area $C'\,3\,D'$ + area $D'\,4\,E'$,

which is equal to the area of the polygon $A'\,1\,2\,3\,4\,E'$. Hence this area, multiplied by $2H$, gives the moment of inertia of the required system of forces.

It may sometimes be convenient to apply this principle in determining moments of inertia, the area being determined by use of a planimeter, or by any other convenient method. It should be noticed that if H is taken equal to the sum of the given forces (AE), twice the area of the funicular polygon is equal to the square of the radius of gyration. If $H = \tfrac{1}{2} AE$, the

GRAPHIC STATICS.

square of the radius of gyration is equal to the area of the polygon.

194. Determination of Product of Inertia of Parallel Forces. — Assume the points of application of the forces to be in the plane containing the two axes. If the moment of any force with respect to one axis be found, and a force equal in magnitude to this moment be assumed to act at the point of application of the original force, then the moment of this new force with respect to the second axis is equal to the product of inertia of the given force for the two axes.

Thus, let *ab, bc, cd, de* (Fig. 66) be the points of application of four parallel forces, and let their product of inertia with respect to the axes *QR, ST* be required. Draw *ABCDE*, the

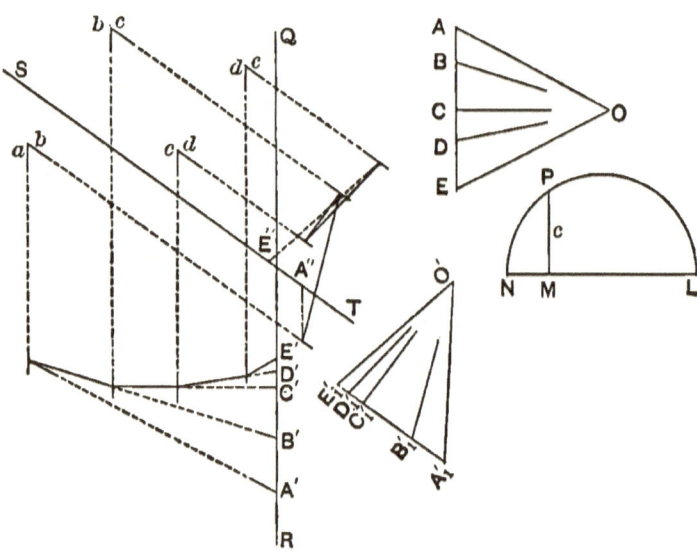

Fig. 66

force polygon for the given forces, assumed to act parallel to *QR*. Choose a pole *O*, the pole distance being preferably taken equal to *AE*, or some simple multiple of *AE*, and draw the funicular polygon as shown, prolonging the strings to intersect *QR* in the points *A', B', C', D', E'*. Now assume a series of forces equal to $A'B'$, $B'C'$, etc., each multiplied by H, to act at the points ab, bc, etc., and determine their moments with respect to the axis ST. To find these moments, draw lines through ab, bc, etc., parallel to ST, and draw a funicular polygon for the assumed forces taken to act in these lines. The force polygon is obtained by revolving the line $A'B'C'D'E'$ until parallel with ST, and is the line $A'_1 B'_1 C'_1 D'_1 E'_1$ in the figure. The strings $o'a'$, $o'c'$ of the second funicular polygon intersect ST in the points A'' and E''. Hence, calling H' the second pole distance, the product of inertia of the given system is equal to $H \times H' \times A''E''$.

195. **Product-Radius.** — If H be taken equal to AE (Fig. 66), $H' \times A''E''$ is equal to the square of the product-radius (Art. 189). Hence the product-radius can be found by a construction exactly like that employed in finding the radius of gyration. Thus (Fig. 66) take $LM = H'$ and $MN = A''E''$; make LN the diameter of a semicircle, and draw from M a line perpendicular to LN, intersecting the semicircle in P. Then $\overline{MP}^2 = LM \times MN = H' \times A''E''$; hence $MP = c$, the product-radius.

196. **Relation between Moments of Inertia for Parallel Axes.** — *Proposition.* — The moment of inertia of a system of parallel forces with reference to any axis is equal to its moment of inertia with respect to a parallel axis through the centroid of the system plus the moment of inertia with respect to the given axis of the resultant applied at the centroid of the system.

Let P_1, P_2, etc., represent the forces; x_1, x_2, etc., the distances of their points of application from the central axis; and a the distance of this central axis from the given axis. Calling the required moment of inertia A, and the moment of inertia with respect to the axis through the center of gravity A', we have

$$A = P_1 (a+x_1)^2 + P_2 (a+x_2)^2 + \cdots$$
$$= a^2 (P_1 + P_2 + \cdots) + 2a (P_1 x_1 + P_2 x_2 + \cdots) + P_1 x_1^2 + P_2 x_2^2 + \cdots$$

Now $P_1x_1+P_2x_2+\cdots$ is the algebraic sum of the moments of the given forces with respect to the axis through their centroid, and is equal to zero; and $P_1x_1^2+P_2x_2^2+\cdots=A'$. Hence

$$A = A' + (P_1+P_2+\cdots)\,a^2,$$

which proves the proposition.

Radii of gyration. — Let k = radius of gyration of the system with respect to the given axis, and k' the radius of gyration with respect to the central axis, and we have

$$A = (P_1+P_2+\cdots)\,k^2.$$
$$A' = (P_1+P_2+\cdots)\,k'^2.$$

Hence the equation above deduced may be reduced to the form

$$k^2 = k'^2 + a^2.$$

197. Products of Inertia with Respect to Parallel Axes. — *Proposition.* — The product of inertia of a system of parallel forces with reference to any two axes is equal to the product of inertia with reference to a pair of central axes parallel to the given axes, plus the product of inertia of the resultant (acting at the centroid) with reference to the given axes.

Let P_1, P_2, etc., be the magnitudes of the given forces; (p_1, q_1), (p_2, q_2), etc., the distances of their points of application from the central axes parallel to the two given axes; (a, b) the distances of the centroid of the system from the given axes. Let A and A' be the products of inertia of the system with respect to the given axes and the parallel central axes respectively. Then

$$A' = P_1 p_1 q_1 + P_2 p_2 q_2 + \cdots$$
$$A = P_1(p_1+a)(q_1+b) + P_2(p_2+a)(q_2+b) + \cdots$$
$$= (P_1 p_1 q_1 + P_2 p_2 q_2 + \cdots) + a(P_1 q_1 + P_2 q_2 + \cdots)$$
$$+ b(P_1 p_1 + P_2 p_2 + \cdots) + ab(P_1 + P_2 + \cdots).$$

But $P_1 q_1 + P_2 q_2 + \cdots = 0$; and $P_1 p_1 + P_2 p_2 + \cdots = 0$; since each of these expressions is the sum of the moments of the given forces

with respect to an axis through the centroid of the system. Hence,
$$A = A' + (P_1 + P_2 + \cdots)ab,$$
which proves the proposition.

If $A = (P_1 + P_2 + \cdots)c^2$ and $A' = (P_1 + P_2 + \cdots)c'^2$, we have
$$c^2 = c'^2 + ab.$$

From the above proposition it follows that if the axes have such directions that the product of inertia with reference to the central axes is zero, the product of inertia with reference to the given axes is the same as if the forces all acted at the centroid. When this condition is known to be satisfied, then for the purpose of finding the product of inertia the system of forces may be replaced by their resultant.

It follows also, in the case when the product of inertia for the central axes is zero, that if one of the given axes coincides with the parallel central axis, the product of inertia for the given axes is zero; for in this case either a or b is zero, and hence $ab(P_1 + P_2 + \cdots)$ is zero. Therefore,

If the product of inertia of a system is zero for two axes, A' and A'', one of which (as A') contains the centroid of a system, then the product of inertia is also zero for A' and any axis parallel to A''.

§ 2. *Moments of Inertia of Plane Areas.*

198. Elementary Areas Treated as Forces. — If any area be divided into small elements, and a force be applied at the centroid of each element numerically equal to its area, the moment of inertia of this system of forces will be approximately equal to that of the given area. The approximation will be closer the smaller the elementary areas are taken. If the elements be made smaller and smaller, so that the area of each approaches zero as a limit, the moment of inertia of the supposed system of forces approaches as a limit the true value of the moment of inertia of the given area.

It is seen, then, that most of the general principles which have been stated regarding moments of inertia of systems of forces are equally applicable to moments of inertia of areas. The practical application of these principles, however, and especially the graphic constructions based upon them, are less simple in the case of areas than of systems of forces such as those already treated. The reason for this is that the system of forces which may be conceived to replace the elements of area consists of an infinite number of infinitely small forces, with which the graphic methods thus far discussed cannot readily deal. Problems of this class are most easily treated by means of the integral calculus, especially when the areas dealt with are in the form of geometrical figures. It is possible, however, by graphic methods to determine approximately the moment of inertia of any plane area; and in many cases exact graphic solutions of such problems are not difficult. The proof of these methods is often most easily effected algebraically.

199. **Moments of Inertia of Geometrical Figures.** — The application of the integral calculus to the determination of moments of inertia will not be here treated. But the values of the moments of inertia of some of the common geometrical figures are of such frequent use that the more important of them will be given for future reference. The moment of inertia is in each case taken with respect to a central axis, and will be represented by I, while the radius of gyration will be called k.

Rectangle. — Let b and d be the sides, the axis being parallel to the side b. Then

$$I = \frac{bd^3}{12}; \quad k^2 = \frac{d^2}{12}.$$

Triangle. — Let b and d be the base and altitude. Then for an axis parallel to the base,

$$I = \frac{bd^3}{36}; \quad k^2 = \frac{d^2}{18}.$$

For an axis through the vertex, bisecting the base, $k^2 = \dfrac{b'^2}{24}$, where b' is the projection of the base on a line perpendicular to the axis.

Circle. — Let d be the diameter. Then
$$I = \frac{\pi d^4}{64}; \quad k^2 = \frac{d^2}{16}.$$

For a central axis perpendicular to the plane of the circle,
$$I = \frac{\pi d^4}{32}; \quad k^2 = \frac{d^2}{8}.$$

Ellipse. — Let a and b be the semi-axes. Then for an axis parallel to a,
$$I = \frac{\pi a b^3}{4}; \quad k^2 = \frac{b^2}{4}.$$

For an axis parallel to b,
$$I = \frac{\pi a^3 b}{4}; \quad k^2 = \frac{a^2}{4}.$$

For a central axis perpendicular to the plane of the ellipse,
$$I = \frac{\pi a b (a^2 + b^2)}{4}; \quad k^2 = \frac{a^2 + b^2}{4}.$$

Graphic construction for radius of gyration. — Whenever k^2 can be expressed as the product of two known factors, the value of k can be found by the construction already used in Art. 191. Thus, in case of a rectangle, for which $k^2 = \dfrac{d^2}{12}$, we may put $k^2 = \dfrac{d}{3} \cdot \dfrac{d}{4}$. Then if in Fig. 65 we take $LM = \dfrac{d}{4}$, $MN = \dfrac{d}{3}$, the construction there shown will give MP as the value of k. For the triangle, the axis being parallel to the base, we have $k^2 = \dfrac{d}{3} \cdot \dfrac{d}{6}$, and the same construction is applicable. For the axis through the vertex bisecting the base, $k^2 = \dfrac{b'}{4} \cdot \dfrac{b'}{6}$.

200. **Product of Inertia.** — *General principles.* — Products of inertia of areas are determined by means of the integral calculus in a manner similar to that employed for moments of inertia. The following fundamental principles regarding products of inertia of geometrical figures will be found useful:

(1) With reference to two *rectangular* axes, one of which is an axis of symmetry (Art. 179), the product of inertia is zero. For it is manifest that the products of inertia of two equal elements, symmetrically placed with reference to one of the axes, are numerically equal but of opposite sign. Hence, if the whole area can be made up of such pairs of elements, the total product of inertia is zero.

(2) If the two axes are not rectangular, but the area can be divided into elements such that for every element whose distances from the axes are p, q, there is an equal element whose distances are p, $-q$, or $-p$, q, the product of inertia is zero. This includes the preceding as a special case.

201. **Products of Inertia of Geometrical Figures.** — In each of the following cases the product of inertia is zero:

A triangle, one axis containing the vertex and the middle point of the base, the other being any line parallel to the base.

A parallelogram, the axes being parallel to the sides and one axis being central. This includes the rectangle as a special case.

An ellipse, the axes being parallel to a pair of conjugate diameters, and one axis being central. This includes, as a special case, that in which one axis is a principal diameter and the other is any line perpendicular to it; and under this case falls also the circle.

202. **Approximate Method for Finding Moment of Inertia of Any Area.** — To apply the method of Art. 190 to the determination of the moment of inertia of a plane area, we should strictly need to replace the area by an infinite number of parallel forces, proportional to the infinitesimal elements of the given area, and

MOMENTS OF INERTIA OF PLANE AREAS. 181

with points of application in these elements. If, instead, we divide the given figure into finite portions whose several areas are known, and assume forces proportional to those areas to act at their centroids, we may get an approximate value for the moment of inertia, which will be more nearly correct the smaller the elements. This will be illustrated by the area shown in Fig. 67.

Let QR be the axis with reference to which the moment of inertia is to be found, in this case taken as a central axis.

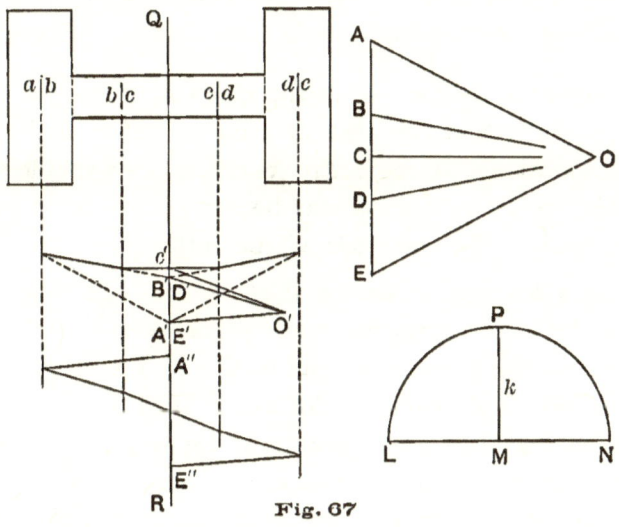

Fig. 67

Divide the figure into four rectangular areas as shown, and assume forces numerically equal to these areas to act at their centroids parallel to QR. The force polygon is $ABCDE$. Draw the funicular polygon corresponding to a pole O, and let the successive strings intersect QR in A', B', C', D', E'. Take this as a new force polygon, and with any convenient pole distance draw a second funicular polygon, using the same lines of action. Let the first and last strings intersect QR in A'' and E''. Then $A''E''$ multiplied by the product of the two pole distances gives the moment of inertia of the four assumed forces, and approximately the moment of inertia of the given figure. If the first pole distance is taken equal to AE (as is the case in Fig. 67), the radius of gyration may be found by the construction of Art. 191. Thus in Fig. 67, MP is the radius of gyration as determined by this method.

A more accurate result may be reached by dividing the area into narrower strips by lines parallel to QR; since the narrower such a strip is, the more nearly will the distance of each small element from the axis coincide with that of the centroid of the strip. If the partial areas are taken as narrow strips of equal width, the forces may be taken proportional to the average lengths of the several strips.

203. **Accurate Methods.** — If the given figure can be divided into parts, such that the area of each is known, and also its radius of gyration with respect to its central axis parallel to the given axis, the above method may be so modified as to give an accurate result. Two methods will be noticed.

(1) *When the axis is known at the start.* — Let the line of action of the force representing any partial area be taken at a distance from the given axis equal to the radius of gyration of that area with reference to the axis. If this is done, it is evident that the moment of inertia of the system of forces is identical with that of the given area. When the axis is known, the position of the line of action for any force may be found as

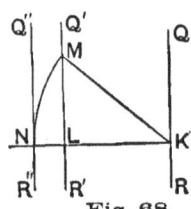
Fig. 68

follows: Let QR (Fig. 68) be the given axis, and $Q'R'$ a parallel axis through the centroid of any partial area. Draw KL perpendicular to QR, and lay off LM equal to the radius of gyration of the partial area with respect to $Q'R'$. Then KM is the length of the radius of gyration with respect to QR. (Art. 196.) Take KN equal to KM and draw $Q''R''$ through N parallel to $Q'R'$; then $Q''R''$ is to be taken as the line of action of the force representing the partial area in question.

(2) *When the axis is at first unknown.* — The method to be

MOMENTS OF INERTIA OF PLANE AREAS.

employed in this case is to let the force representing any partial area act in a line through the centroid of that area; and then assume the force representing its moment to act in such a line that the moment of this second force shall be numerically equal to the moment of inertia of the partial area. This line may be found as follows: Let k represent the radius of gyration of the partial area with respect to its central axis parallel to the given axis, and a the distance between the two axes. Then the moment of inertia of the partial area with respect to the given axis is $A(a^2+k^2)$, if A represents the area. But $A(a^2+k^2) = Aa\left(a+\dfrac{k^2}{a}\right)$. Hence, if a force numerically equal to A is assumed to act with an arm a, then a force equal to its moment Aa must act with an arm $a+\dfrac{k^2}{a}$ in order that its moment may equal $A(a^2+k^2)$.

The distance $a+\dfrac{k^2}{a}$ can be found by a simple construction. Let QR and $Q'R'$ (Fig. 69) be the given axis and the central axis respectively. Draw KL perpendicular to QR and lay off LM equal to k. From M draw a line perpendicular to KM, intersecting KL produced at N. Then $KN = a+\dfrac{k^2}{a}$. For, in the similar triangles KNM, KML, we have

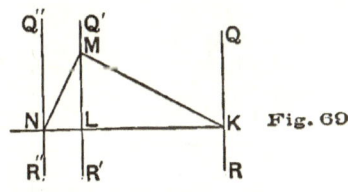

Fig. 69.

$$\frac{KN}{KM} = \frac{KM}{KL}; \text{ or } KN = \frac{\overline{KM}^2}{KL}.$$

But $KL = a$, and $\overline{KM}^2 = a^2+k^2$; hence

$$KN = \frac{a^2+k^2}{a} = a+\frac{k^2}{a}.$$

This second method is more useful than the first, because in applying it the first funicular polygon can be drawn before the position of the inertia-axis is known. Thus, a very common

184   GRAPHIC STATICS.

case is that in which the moment of inertia of an area is to be found for a central axis, whose direction is known, while at the outset its position is unknown because the centroid of the area is unknown. If the second method be employed, the first funicular polygon can be drawn at once, and serves to locate the required central axis, as well as to determine the moments of the first set of forces as soon as the axis is known. The central axis and the moment of inertia with respect to it are thus determined by a single construction.

Example. — The method last described is illustrated in Fig. 70. The area shown consists of two rectangles, the centroids of which are marked *ab* and *bc*. The moment of inertia is to be found for a central axis parallel to the longer side of the rectangle *ab*.

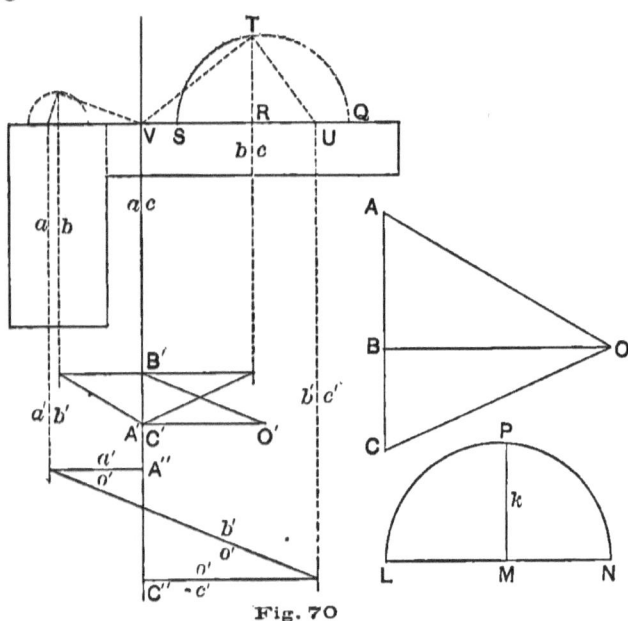

Fig. 70

We draw through *ab* and *bc* lines parallel to the assumed direction of the axis, and take these for the lines of action of forces *AB*, *BC*, proportional to the areas of the two rectangles. *ABC* is the force polygon, and the pole distance is taken equal to AC. The intersection of the strings oa, oc, determines a point of the required central axis ac. The moments of the given forces are proportional to $A'B'$, $B'C'$. We now have to find the lines of action for the forces $A'B'$, $B'C'$, in accordance with the method above described.

Take any line perpendicular to the axis ac, as VQ, the side of one of the rectangles. From R, the point in which this line intersects the vertical line through bc, lay off RT equal to the central radius of gyration of the rectangle whose centroid is bc. To find this central radius of gyration, we know that its value is $\sqrt{\frac{d^2}{12}}$ (Art. 199), where d is the length of the side of the rectangle perpendicular to the axis. Hence we take $QR = \frac{d}{3}$, $RS = \frac{d}{4}$, and make QS the diameter of a semicircle, intersecting the vertical line bc in T; then RT is the required radius of gyration of the rectangle with respect to a central axis. Now draw from T a line perpendicular to VT, intersecting VQ in U; then the line $b'c'$ drawn through U parallel to the given axis is the line of action of the force $B'C'$.

By a similar construction applied to the other rectangle, $a'b'$ is located as the line of action of the force $A'B'$. The second funicular polygon is now drawn, and the points A'', C'' are found by the intersection of the strings $o'a'$, $o'c'$ with the axis. Hence the moment of inertia of the given area is equal to $A''C'' \times H \times H'$. In the figure H is made equal to AC, hence the radius of gyration can be determined by the usual construction, and its length is found to be MP.

204. Moment of Inertia of Area Determined from Area of Funicular Polygon. — The method given in Art. 193 for finding the moment of inertia of a system of forces by means of the area of the funicular polygon may be applied with approximate results to the case of a plane figure. If the forces are taken as acting at the centroids of the areas they represent, then to get good results these partial areas should be taken as narrow strips between lines parallel to the axis.

If the lines of action are determined as in the first method of the preceding article, then the area enclosed by the funicular polygon and the axis represents accurately the required moment of inertia.

205. Product of Inertia Determined Graphically. — To determine the product of inertia of any area, let it be divided into small known parts, and let parallel forces numerically equal to the partial areas be assumed to act at the centroids of these parts. The product of inertia of these forces may then be found as in Art. 194, and its value will represent approximately the product of inertia of the given area.

If the partial areas can be so taken that the product of inertia of each with reference to axes through its centroid parallel to the given axes is zero, the method here given is exact. (Art. 197.)

If the partial areas are taken as narrow strips parallel to one of the axes, the condition just mentioned will be nearly fulfilled; for the product of inertia of each strip for a pair of axes through its centroid, one of which is parallel to its length, will be very small.

CHAPTER X. CURVES OF INERTIA.

§ 1. *General Principles.*

206. Relation between Moments of Inertia for Different Axes through the Same Point. — The moment of inertia with respect to any axis through a given point can be expressed in terms of the moments and product of inertia for any two axes through that point. It is necessary here to use algebraic methods, but the results reached form the basis of graphic constructions.

Let OX, OY (Fig. 71) be the two given axes; θ, the angle included between them; P_1, P_2, etc., the forces of the system; x', y', the coördinates of the point of application of any force P, referred to the axes OX, OY; p, q, the perpendicular distances of the same point from OY and OX respectively, so that

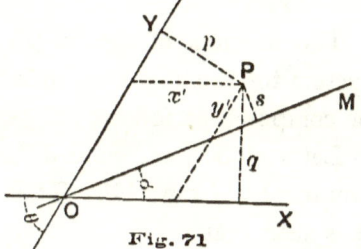

Fig. 71

$$p = x' \sin \theta, \quad q = y' \sin \theta.$$

Let a', b', and c' be quantities defined by the equations

$$a'^2 = \frac{P_1 x_1'^2 + P_2 x_2'^2 + \cdots}{P_1 + P_2 + \cdots} = \frac{\Sigma P x'^2}{\Sigma P},$$

$$b'^2 = \frac{\Sigma P y'^2}{\Sigma P},$$

$$c'^2 = \frac{\Sigma P x' y'}{\Sigma P}$$

Then
$$a'^2 \sin^2 \theta = \frac{\Sigma P x'^2 \sin^2 \theta}{\Sigma P} = \frac{\Sigma P p^2}{\Sigma P},$$

$$b'^2 \sin^2 \theta = \frac{\Sigma P y'^2 \sin^2 \theta}{\Sigma P} = \frac{\Sigma P q^2}{\Sigma P},$$

$$c'^2 \sin^2 \theta = \frac{\Sigma P x'y' \sin^2 \theta}{\Sigma P} = \frac{\Sigma P pq}{\Sigma P};$$

and $a' \sin \theta$, $b' \sin \theta$, $c' \sin \theta$ are respectively the radius of gyration with respect to OY, the radius of gyration with respect to OX, and the product-radius (Art. 189) with respect to OX and OY.

The moment of inertia (I) and radius of gyration (k) of the system for the axis OM, making an angle ϕ with OX, may now be computed as follows:

Let $s = $ perpendicular distance of the point of application of any force P from OM. Then from the geometry of the figure it is seen that
$$s = y' \sin(\theta - \phi) - x' \sin \phi.$$
Hence
$$I = \Sigma P s^2 = \Sigma P y'^2 \sin^2(\theta - \phi) - 2 \Sigma P x'y' \sin(\theta - \phi) \sin \phi$$
$$+ \Sigma P x'^2 \sin^2 \phi;$$
or,
$$I = b'^2 \Sigma P \cdot \sin^2(\theta - \phi) - 2 c'^2 \Sigma P \cdot \sin(\theta - \phi) \sin \phi$$
$$+ a'^2 \Sigma P \cdot \sin^2 \phi,$$

the factors involving θ and ϕ being constant for all terms of the summation. Hence

$$k^2 = \frac{I}{\Sigma P} = b'^2 \sin^2(\theta - \phi) - 2 c'^2 \sin(\theta - \phi) \sin \phi + a'^2 \sin^2 \phi \quad . . \quad (1)$$

From these equations I and k may be computed if a', b', and c' are known; that is, if the moments and product of inertia for the two axes OX and OY are known.

Special case. — If $\theta = 90°$, the equation (1) becomes
$$k^2 = b'^2 \cos^2 \phi - 2 c'^2 \sin \phi \cos \phi + a'^2 \sin^2 \phi \quad . . . . (2)$$

207. Products of Inertia for Different Axes through the Origin. — The product of inertia with respect to OM and OX may be found as follows:

Let A = the required product of inertia; then

$$A = \Sigma Pqs = \Sigma Py' \sin\theta \, [y' \sin(\theta-\phi) - x' \sin\phi]$$
$$= \Sigma P \, [y'^2 \sin\theta \sin(\theta-\phi) - x'y' \sin\theta \sin\phi]$$
$$= [b'^2 \sin\theta \sin(\theta-\phi) - c'^2 \sin\theta \sin\phi] \, \Sigma P.$$

Let h = product-radius for axes OM and OX; then

$$h^2 = \frac{A}{\Sigma P} = b'^2 \sin\theta \sin(\theta-\phi) - c'^2 \sin\theta \sin\phi.$$

Special case. — The axis OM may be so chosen that $A = 0$. This will be the case if

$$b'^2 \sin(\theta-\phi) = c'^2 \sin\phi.$$

208. Inertia Curve. — If on OM (Fig. 71) a point M be taken such that the length OM depends in some given way upon the value of k, and if similar points be located for all possible directions of OM, the locus of such points will be a curve which is called a *curve of inertia* of the system for the center O.

The form of the curve will depend upon the assumed law connecting OM with k.

209. Ellipse or Hyperbola of Inertia. — The simplest curve is obtained by assuming OM to be inversely proportional to k. Let $OM = r$, and take $r^2 = \dfrac{d^4 \sin^2\theta}{k^2}$, where d^2 is a positive quantity, so that d always represents a *real* length, positive or negative.

Equation (1) of Art. 206 then becomes

$$d^4 \sin^2\theta = b'^2 r^2 \sin^2(\theta-\phi) - 2c'^2 r^2 \sin(\theta-\phi)\sin\phi + a'^2 r^2 \sin^2\phi,$$

which is the polar equation of the inertia-curve, r and ϕ being the variable coördinates. Let x, y be coördinates of the point M referred to the axes OX, OY. Then

$$\frac{r}{\sin\theta} = \frac{x}{\sin(\theta-\phi)} = \frac{y}{\sin\phi},$$

whence
$$\frac{r^2 \sin^2(\theta-\phi)}{\sin^2\theta} = x^2,$$

$$\frac{r^2 \sin^2\phi}{\sin^2\theta} = y^2,$$

$$\frac{r^2 \sin(\theta-\phi)\sin\phi}{\sin^2\theta} = xy,$$

and the equation becomes

$$d^4 = b'^2 x^2 - 2c'^2 xy + a'^2 y^2 \quad \ldots \ldots \ldots \quad (3)$$

The form of this equation shows that it represents a conic section whose center is at the origin of coördinates O. This conic may be either an ellipse or a hyperbola.

The equation will be discussed more fully in a later article. One fact may, however, be here noticed.

If the moment of inertia I is positive for all positions of the axis, the radius of gyration k will be real, whatever the value of ϕ. But r, the radius vector of the curve, will be real when k is real; hence in this case the curve is an ellipse. This is always the case if the given forces have all the same sign.

If the forces have not all the same sign, it is possible that the value of I may have different signs for different directions of the axis. If this is so, certain values of ϕ make k (and therefore r) imaginary. In this case the curve is a hyperbola.

The most important case is that in which the given forces have all the same sign which may be taken as plus, so that the moment of inertia is always positive, and the curve an ellipse; and to this case the discussion will be confined.

§ 2. *Inertia-Ellipses for Systems of Forces.*

210. Properties of the Ellipse. — In discussing ellipses of inertia use will be made of certain general properties of the ellipse, which, for convenience of reference, will be here summarized. For the proof of the propositions stated the reader is referred to works on the conic sections.

INERTIA-ELLIPSES FOR SYSTEMS OF FORCES.

(1) The equation
$$Ax^2 + 2Bxy + Cy^2 = D$$
represents an ellipse if $B^2 - AC$ is negative; a hyperbola if $B^2 - AC$ is positive. (Salmon's *Conic Sections*, p. 140.) The coördinate axes may be either rectangular or oblique.

(2) Two diameters of an ellipse are said to be *conjugate* to each other if each bisects all chords parallel to the other. If the axes of coördinates coincide with a pair of conjugate diameters, the lengths of which are $2a'$ and $2b'$, the equation of the curve is
$$\frac{x^2}{a'^2} + \frac{y^2}{b'^2} = 1.$$

A particular case of this equation is that in which the coördinate axes are rectangular, being the principal axes of the curve; in which case we may write a and b instead of a' and b'.

(3) In an ellipse, the product of any semi-diameter and the perpendicular from the center on the tangent parallel to that semi-diameter is constant and equal to ab. That is, if r is any radius vector of the curve drawn from the center, and p the length of the perpendicular from the center to the parallel tangent, we have
$$pr = ab$$
where a and b are the principal semi-axes of the curve.

(4) Let a' and b' be conjugate semi-diameters. Then each is parallel to the tangent at the extremity of the other. Hence the length of the perpendicular from the center to the tangent parallel to a' is $b' \sin \theta$, where θ is the angle included between a' and b'. Therefore from the preceding paragraph,
$$a'b' \sin \theta = ab.$$

(5) An ellipse can be constructed, when a pair of conjugate diameters is known, as follows:

Let AA', BB' (Fig. 72), be the conjugate diameters, O being the center of the ellipse. Complete the parallelogram $OBCA$.

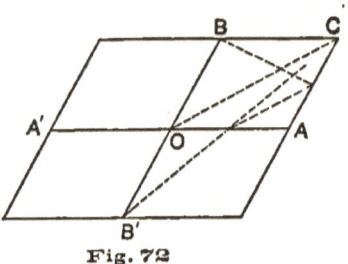

Fig. 72

Divide OA and CA into parts proportional to each other, beginning at O and C. Through the points of division of OA draw lines radiating from B', and through the points of division of CA draw lines radiating from B. The points of intersection of the corresponding lines in the two sets are points of the ellipse. In a similar way, the other three quadrants may be drawn. (The location of one point is shown in the figure.)

A convenient way to locate the corresponding points of division on OA and CA is to cut these lines by lines parallel to the diagonal OC.

211. Discussion of Equation of Inertia-Curve. — We will now examine the equation of the inertia-curve,

$$b'^2 x^2 - 2 c'^2 xy + a'^2 y^2 = d^4$$

with reference to the properties of the ellipse above enumerated.

(1) If $a'^2 b'^2 - c'^4$ is positive, the equation denotes an ellipse. This cannot be the case if a'^2 and b'^2 have opposite signs. But from the definitions of a'^2 and b'^2 (Art. 206) it is seen that their signs are the same as those of the moments of inertia for Y and X axes respectively. Hence, if there are any two axes through the assumed center for which the moments of inertia have opposite signs, the inertia-curve is a hyperbola.

If the moment of inertia has the same sign for all axes through the assumed center, the curve is an ellipse. For, since $c'^2 \sin^2 \theta = \dfrac{\Sigma Ppq}{\Sigma P}$ (Art. 206), c' may be made zero by choosing the axes so that the product of inertia with respect to them is zero; and if c' is zero, and a'^2 and b'^2 have the same sign, the quantity $a'^2 b'^2 - c'^4$ is positive.

This agrees with the conclusion stated in Art. 209.

We shall here deal only with ellipses of inertia.

(2) If $c'=0$, the coördinate axes are conjugate axes of the curve. But the condition $c'=0$ means that the product of inertia for the two axes is zero. Hence any two axes for which the product of inertia is zero are conjugate axes of the inertia-curve. (This is true whether the curve is an ellipse or a hyperbola.)

(3) By the law of formation of the inertia-conic (Art. 209), the length of the radius vector lying in any line is inversely proportional to the radius of gyration with respect to that line. But by (3) of the last article, the perpendicular from the center on the tangent parallel to any radius vector is inversely proportional to the length of that radius vector. Hence the perpendicular distance between any diameter and the parallel tangent is directly proportional to the radius of gyration with respect to that diameter. The curve may be so constructed that the length of this perpendicular is *equal* to the radius of gyration, as follows :

From Art. 209, we have

$$k = \frac{d^2 \sin \theta}{r}$$

and from (3) and (4) of the last article we have

$$p = \frac{ab}{r} = \frac{a'b' \sin \theta}{r},$$

if a' and b' are *conjugate* semi-diameters. Now take

$$d^2 \sin \theta = ab = a'b' \sin \theta,$$

or
$$d^2 = a'b',$$

and we have
$$k = p,$$

and the equation of the curve becomes (since $c'=0$ when the axes are conjugate)

$$b'^2 x^2 + a'^2 y^2 = a'^2 b'^2.$$

If the equation be written in this form, a' and b' having the meanings assigned in Art. 206, *the radius of gyration about any axis through the center of the ellipse is equal to the perpendicular distance between the axis and the parallel tangent to the ellipse.*

Hereafter we shall mean by *inertia-ellipse* the curve obtained by taking $d^2 = a'b'$ as above described, so that the radius of gyration for any axis can be found by direct measurement when a parallel tangent to the ellipse is known.

212. To Determine Tangents to the Inertia-Ellipse for Any Center. — Let the radius of gyration (k) be found for any assumed axis through the given center by one of the methods already described (Arts. 202 and 203). Then two lines parallel to the axis and distant k from it, on opposite sides, will be tangents to the inertia-ellipse.

213. To Construct the Inertia-Ellipse, a Pair of Conjugate Axes Being Known in Position. — If the positions of two axes conjugate to each other can be found, the ellipse can be drawn by the following method:

Determine the radius of gyration for each of the two axes and draw the corresponding tangents as in the preceding article; then proceed as follows:

Let XX', YY' (Fig. 73) be the given axes, and let the four tangents determined as above form the parallelogram $PQRS$.

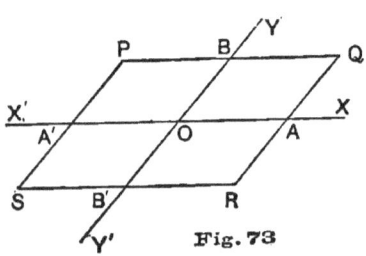

Fig. 73

Let A, A', B, B' be the points in which these tangents intersect the axes XX', YY'. Then, since each diameter is parallel to the tangents at the extremities of the conjugate diameter, A, A', B, B' are the extremities of the diameters lying in the given axes. The ellipse can now be constructed as explained in Art. 210 (Fig. 72).

This method of constructing the inertia-ellipse is useful

whenever the given system has a pair of conjugate axes which can be located by inspection.

214. Central Ellipse. — It is evident that an inertia-curve can be found with its center at any assumed point. That ellipse whose center is the centroid of the given system is called the *central ellipse* for the system.

Since the central ellipse gives at once the radius of gyration for every axis through the centroid of the system, it enables us to determine readily the radius of gyration for any axis whatever, by means of the known relation between radii of gyration for parallel axes. (Art. 196.)

§ 3. *Inertia-Curves for Plane Areas.*

215. General Principles. — The principles deduced in the treatment of inertia-curves for systems of forces are all true for the case of plane areas. But special difficulties arise in dealing with areas, because of the fact that the system of forces equivalent to any area consists of an infinite number of forces. The principles already developed are, however, sufficient to deal at least approximately with all areas, and accurately with many cases.

216. Inertia-Curve an Ellipse. — Since the forces conceived to replace the elements of area (Art. 198) have all the same sign, the value of k^2 is always positive, and the inertia-curve is always an ellipse. (Arts. 209 and 211.)

217. Cases Admitting Simple Treatment. — Whenever a pair of conjugate diameters can be located, and the radius of gyration determined for each, the inertia-ellipse can be at once drawn as in Art. 213. This will be the case whenever it is possible to locate readily a pair of axes for which the product of inertia is zero.

(1) If there is an axis of symmetry, this and any line perpendicular to it are a pair of conjugate axes (and in fact the principal axes) of the inertia-ellipse whose center is at their intersection. (Art. 200.)

(2) If two axes can be located in such a way that for every element of area whose distances from the axes are p, q, there is an equal element whose distances are p, $-q$, or $-p$, q, the product of inertia is zero for the two axes, and these are therefore a pair of conjugate axes of the inertia-ellipse whose center is at their intersection. This of course includes the case when there is an axis of symmetry. (Art. 200.)

When a pair of conjugate axes is known, the radius of gyration is to be found by one of the methods of Art. 202 or Art. 203; the ellipse can then be drawn exactly as explained in Art. 213.

If a pair of conjugate axes cannot be located by inspection, the inertia-ellipse cannot be so readily constructed. Such cases will not be here treated.

As examples of areas, in which the principal axes of the inertia-curve can be located by inspection, may be mentioned the cross-section of the I-beam, the deck-beam, the channel-bar, and other shapes of structural iron and steel.

Many geometrical figures possess axes of symmetry. In others a pair of conjugate axes can be located by principle (2). Some of these will be discussed in the next article.

Example. — Draw the central ellipse for the deck-beam section shown in Fig. 74.

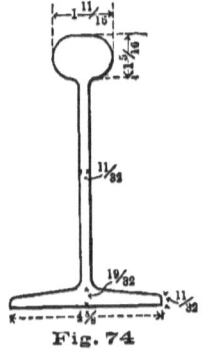

Fig. 74.

[SUGGESTIONS. — Since there is an axis of symmetry, this contains one of the principal axes of the ellipse. The other can be drawn as soon as the centroid of the section is known. Find the radius of gyration for each axis by Art. 202, and then construct the ellipse as explained in Art. 213.]

218. **Central Ellipses for Geometrical Figures.** — In many of the simple geometrical figures, not only can a pair of conjugate axes be located by inspection, but the radius of gyration for each of these axes can be found by a simple construction, so that the central ellipse can be readily drawn. Some of these cases will be here summarized.

INERTIA-CURVES FOR PLANE AREAS. 197

(1) *Parallelogram.* — Let $ABCD$ (Fig. 75) be the parallelogram; then XX', YY', drawn through the centroid parallel to the sides, are a pair of conjugate axes of the central ellipse. Let $AB = b$, $BC = d$, and let $h =$ the perpendicular distance between AB and DC. The moment of inertia of the parallelogram with respect to the axis XX' is equal to the moment of inertia of a rectangle of sides b and h. Hence k^2, the square of the radius of gyration for this axis, is $\dfrac{h^2}{12}$. The length of k can be found by the construction used in case of the rectangles in Fig. 70. The following modification of the method is, however, more convenient:

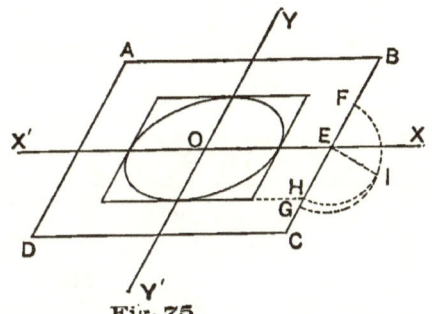

Fig. 75

Make $EF = \frac{1}{4} BC$, $EG = \frac{1}{3} BC$, and draw a semicircle with FG as a diameter. From E draw a line perpendicular to BC, intersecting the semicircle at I. Lay off $EH = EI$; then a line through H parallel to XX' is a tangent to the central ellipse. For by construction,

$$EH = EI = \frac{BC}{\sqrt{12}}.$$

And since the projection of BC on a line perpendicular to XX is equal to h, the projection of EH on the same line is equal to $\dfrac{h}{\sqrt{12}}$, that is to k.

The tangent parallel to the side BC may be found in a similar way. It may, however, be located more simply as follows: It is evident that the distance between YY' and a tangent parallel to it, measured along AB, bears the same ratio to AB that EH does to BC. Hence, the parallelogram formed by the four tangents, two parallel to XX' and two parallel to YY', is similar to the parallelogram $ABCD$.

Fig. 75 shows this parallelogram and also the ellipse.

(2) *Triangle.* — Let ABC (Fig. 76) be the triangle; $b =$ length of the base BC; $b' =$ length of projection of BC on a line perpendicular to AD; $d =$ altitude measured perpendicular to BC. AD and a line through the centroid parallel to BC are a pair of conjugate axes of the central ellipse (Art. 217). From Art. 199, the radius of gyration for a central axis parallel to BC is

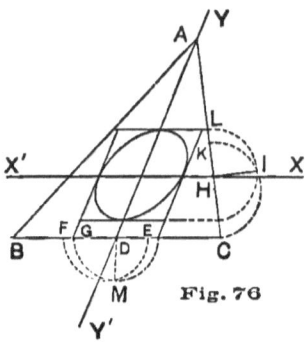

Fig. 76

$\sqrt{\dfrac{d^2}{18}} = \sqrt{\dfrac{d}{3} \cdot \dfrac{d}{6}}$. Let XX' be this axis, and H the point in which it intersects AC. Then $HC = \tfrac{1}{3} AC$. Take $HK = \tfrac{1}{6} AC = \tfrac{1}{2} HC$, and make KC the diameter of a semicircle. From H draw HI perpendicular to AC, intersecting the semicircle at I. Make $HL = HI$; then the line through L parallel to XX' is a tangent to the central ellipse. For the radius of gyration with respect to XX' is to HL as the altitude d is to AC.

Again, for the axis AD, the radius of gyration is $\sqrt{\dfrac{b'^2}{24}}$ (Art. 199). Make $DE = \tfrac{1}{6} BC$ and $DF = \tfrac{1}{4} BC$, and take EF as a diameter of a semicircle. From D draw a line perpendicular to BC, intersecting the semicircle in M; and make $DG = DM$; then a line from G parallel to AD is a tangent to the central ellipse.

The figure shows the parallelogram formed by the two tangents parallel to XX' and the two parallel to AD, and also the central ellipse.

(3) *Ellipse.* — From Art. 199, the radii of gyration of an ellipse with respect to the two principal diameters are $\tfrac{1}{2} a$ and $\tfrac{1}{2} b$. Hence the central ellipse of inertia is similar to the given ellipse, its semi-axes being $\tfrac{1}{2} a$ and $\tfrac{1}{2} b$. A special case of this is a circle, for which the central curve is a circle whose radius is half that of the given circle.

(4) *Semicircle.* — Let ABC (Fig. 77) be the semicircle, O being the centroid. (The point O may be located by the method described in Art. 184.) From symmetry it is evident that the principal axes of the central ellipse are XX' and YY', drawn through O, respectively parallel and perpendicular to AB.

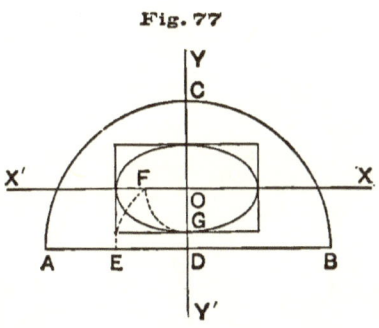

Fig. 77

With respect to the axis YY', the radius of gyration is evidently $\frac{1}{2} r$, the same as for the whole circle. Hence two lines parallel to YY' and distant $\frac{1}{2} r$ from it are tangents to the central ellipse.

Again, the radius of gyration of the semicircle with respect to AB as an axis is also $\frac{1}{2} r$, the same as for the whole circle. To find it for the axis XX', with D as a center, and radius $\frac{1}{2} r$, draw an arc intersecting XX' at F; then $\overline{OF}^2 = \overline{DF}^2 - \overline{OD}^2$. But DF is equal to the radius of gyration with respect to AB, and OD is the distance between XX' and AB; hence (Art. 196) OF is equal to the radius of gyration with respect to XX'. Hence if two lines are drawn parallel to XX', each at a distance from it equal to OF, they will be tangents to the central ellipse. The ellipse can now be drawn in the usual manner.

219. **Summary of Results.** — By the principles and methods developed in the present chapter, inertia-curves can be drawn for all the simpler cases that may arise; namely, whenever a pair of conjugate axes can be located by inspection. This will be the case whenever the product of inertia can be seen to be zero for any pair of axes; and it includes every case of an area possessing an axis of symmetry.

It is believed that this chapter contains as complete a discussion as is needed by the student of engineering. Those who desire to pursue the subject further may consult other works.

Fig. 34

(*A*)

Scale, 1 inch = 6 feet.

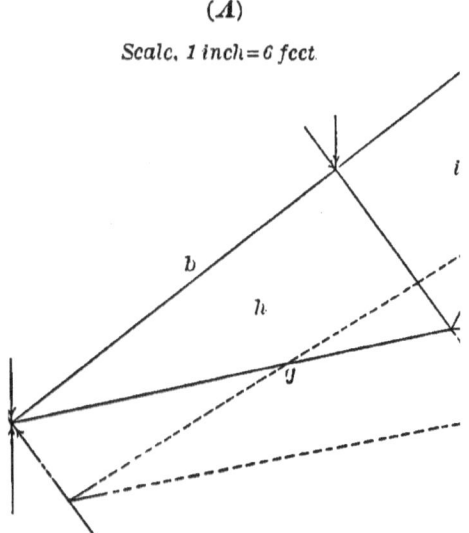

(*B*)

Scale, 1 inch = 1800 lbs.

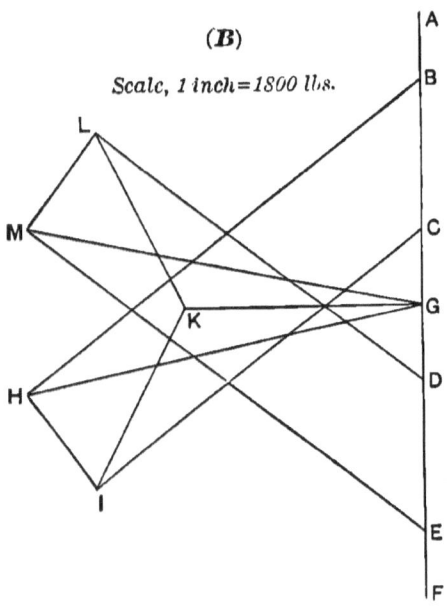

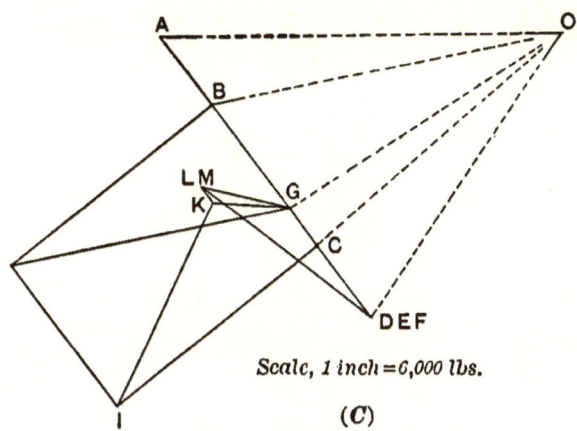

Scale, 1 inch = 6,000 lbs.

(*C*)

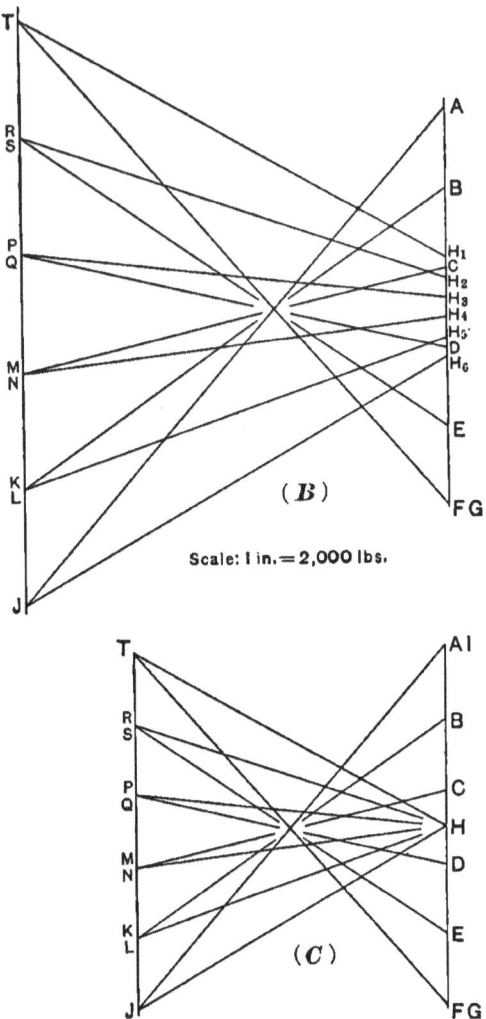

Scale: 1 in. = 2,000 lbs.

(B)

Scale: 1 in. = 4,000 lbs.

(C)

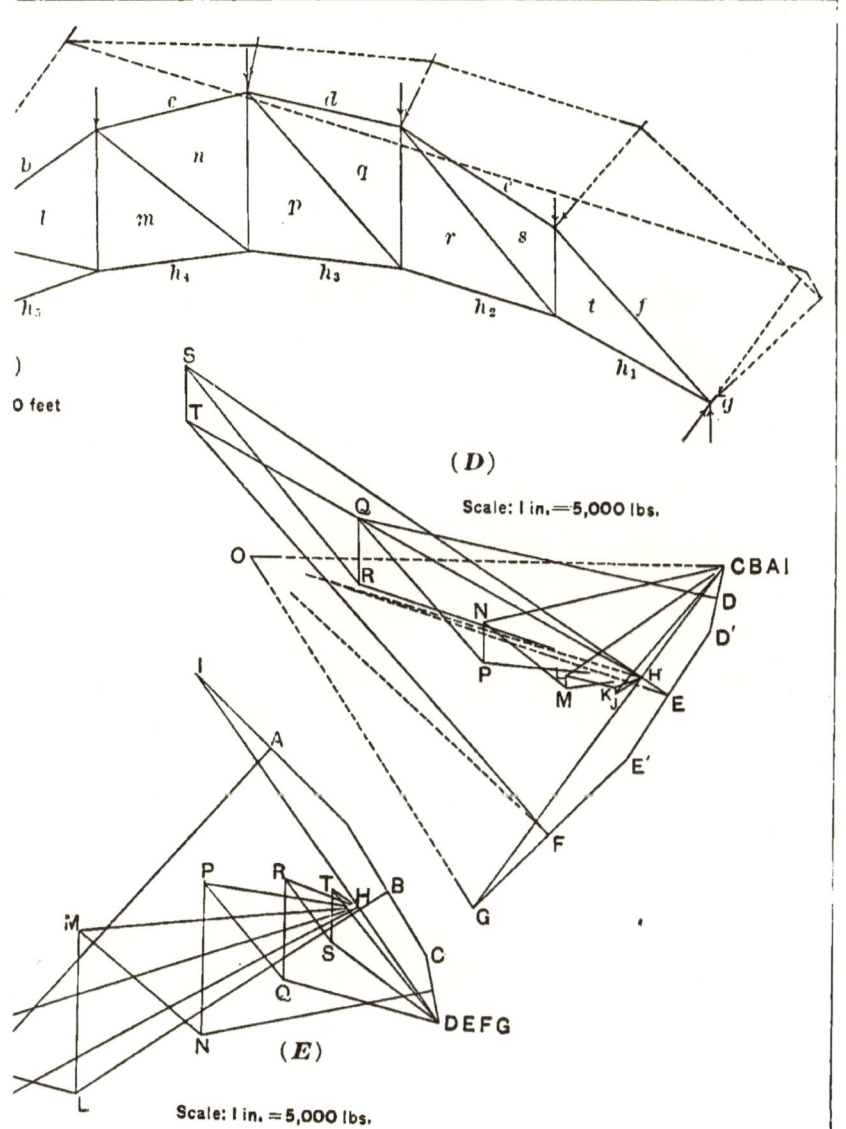

PLATE II.

(D)

Scale: 1 in.=5,000 lbs.

(E)

Scale: 1 in.=5,000 lbs.

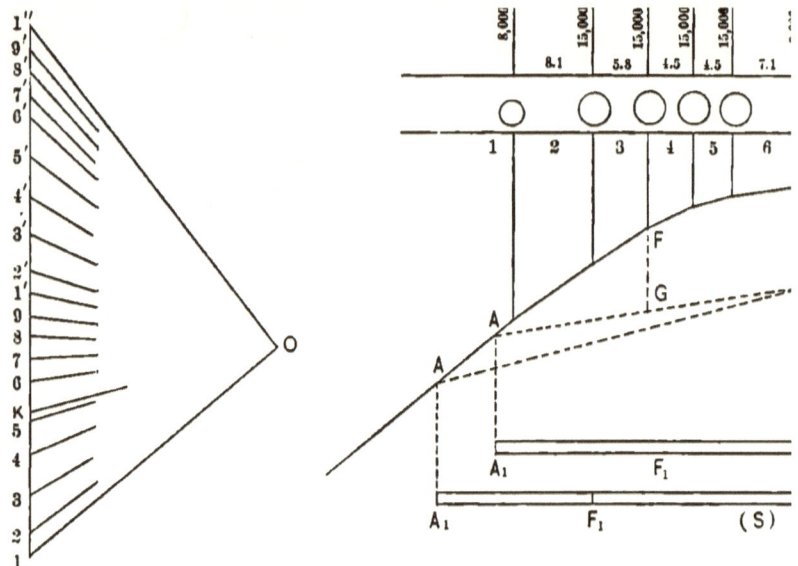

Linear Scale, 1 in. = 20 ft.
Force Scale, 1 in. = 30,000 lbs.

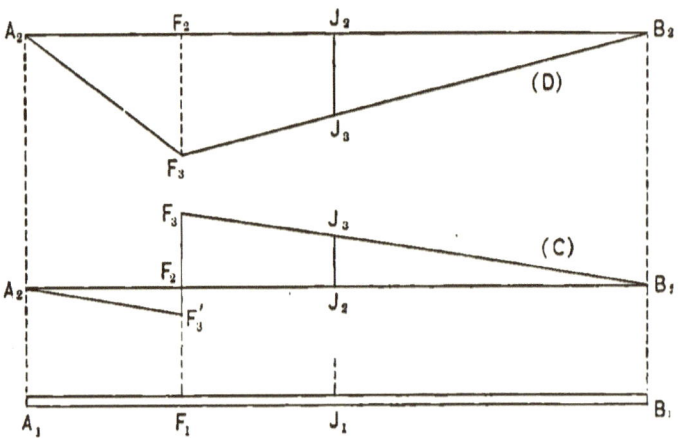

PLATE III.

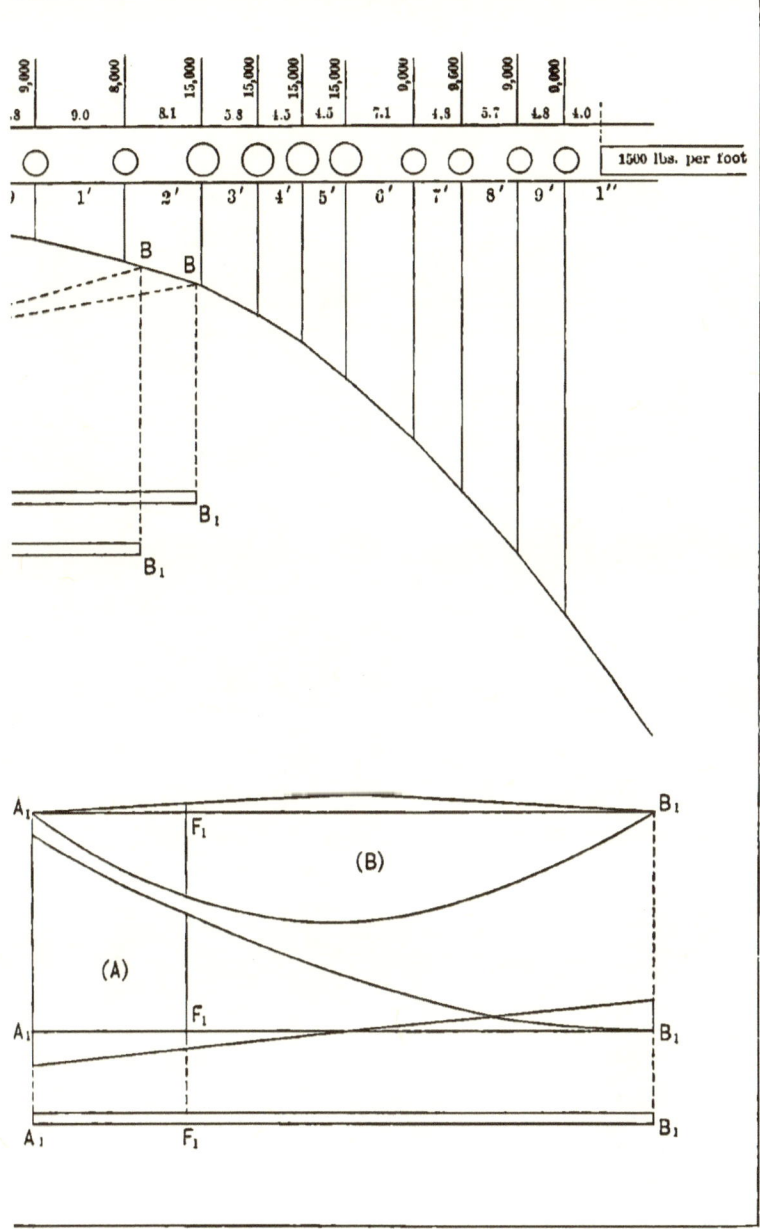

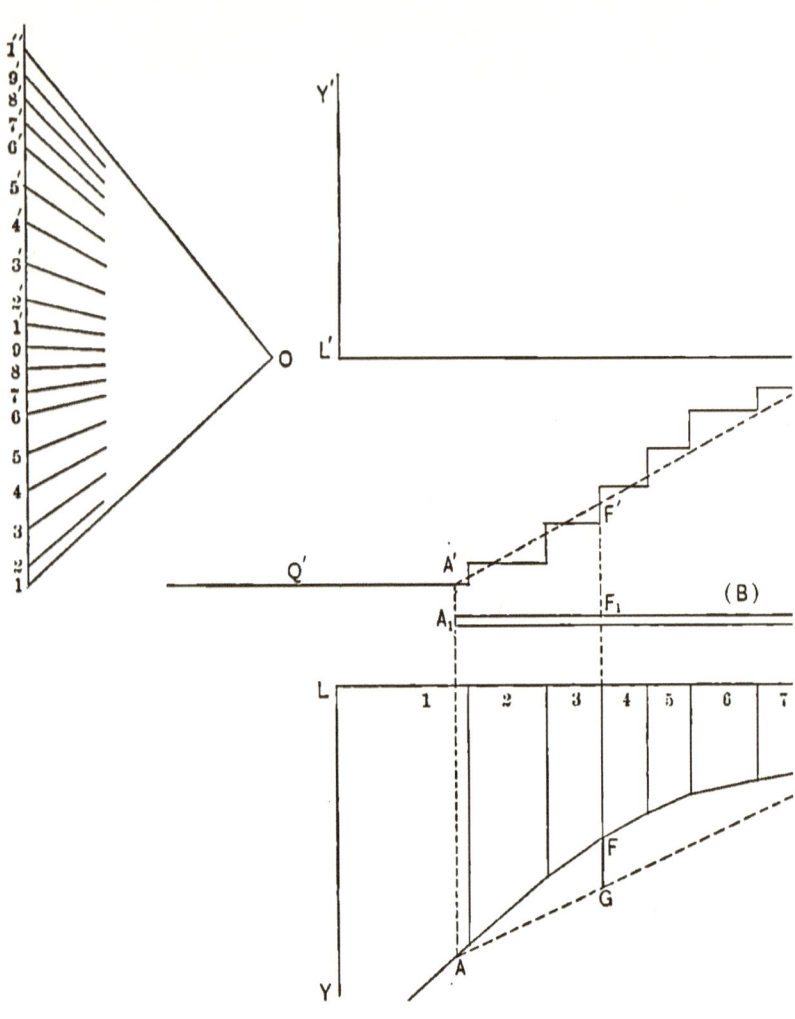

PLATE IV.

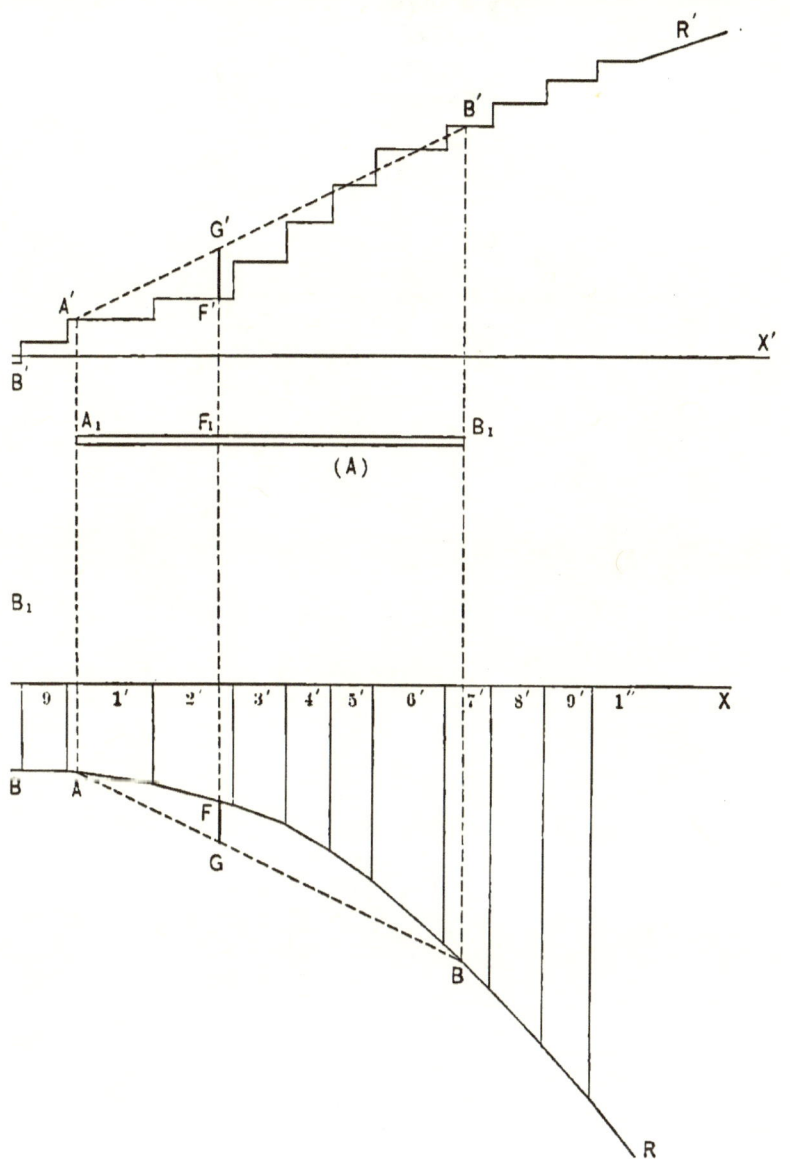

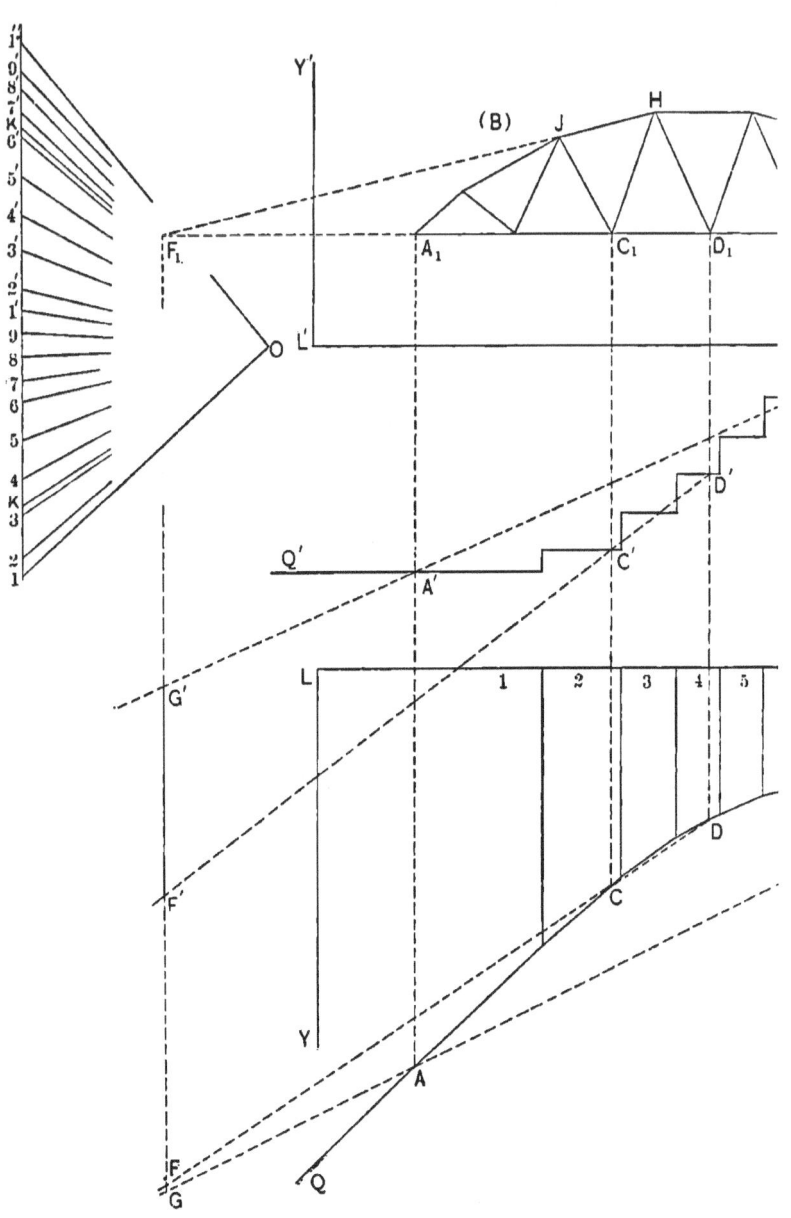

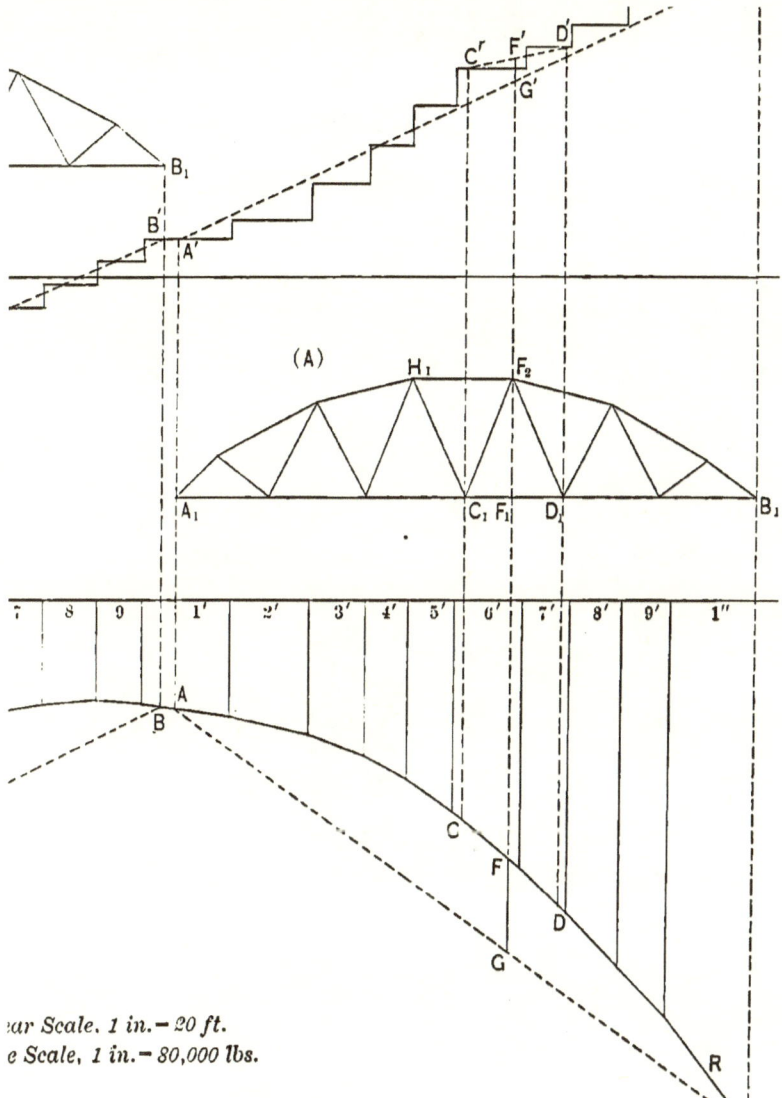

ar Scale. 1 in. = 20 ft.
e Scale, 1 in. = 80,000 lbs.

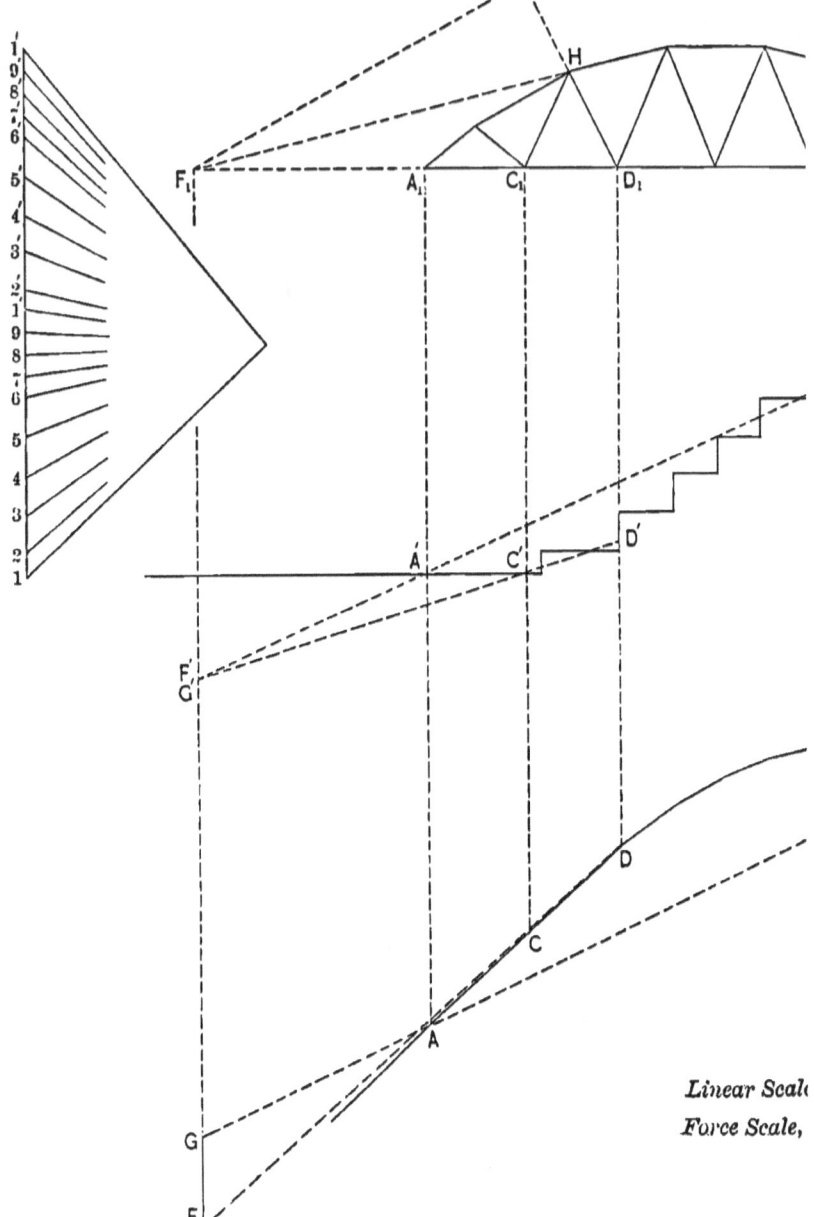

Linear Scale
Force Scale,

PLATE VI.

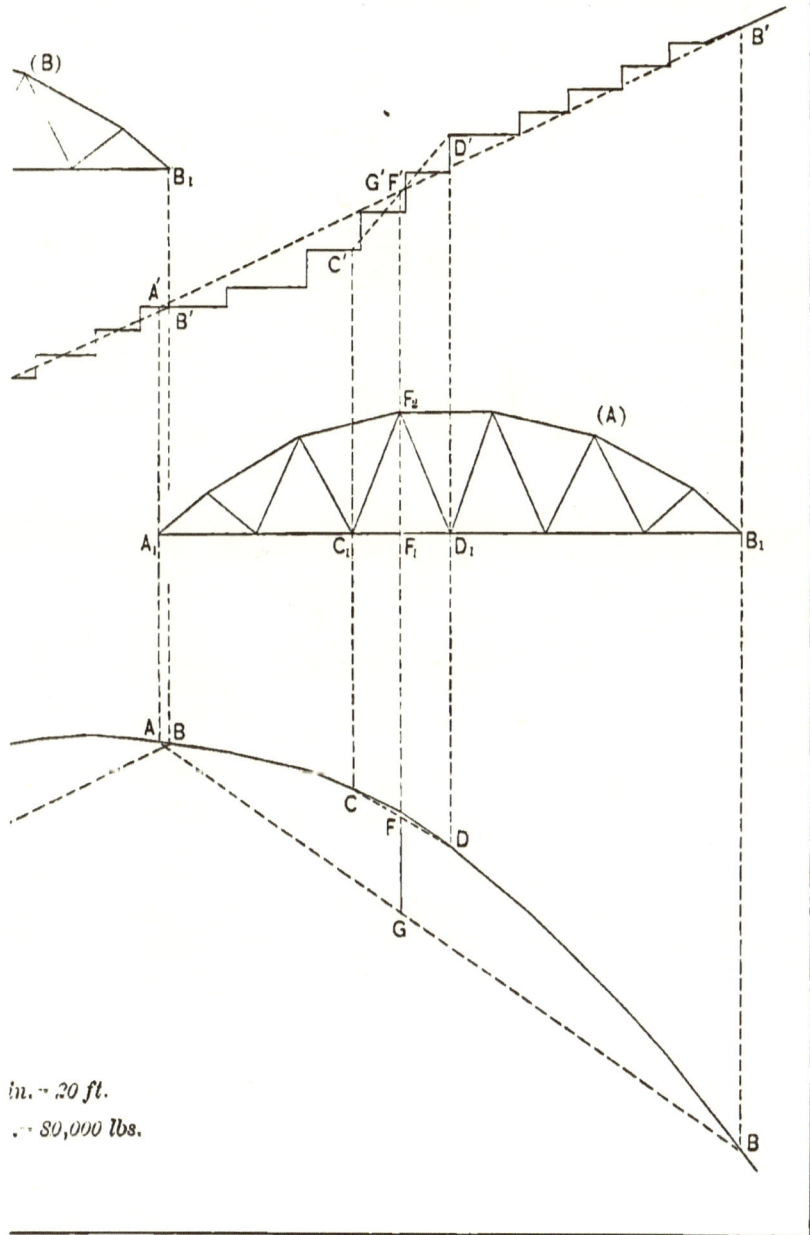

in. = 20 ft.
. = 80,000 lbs.

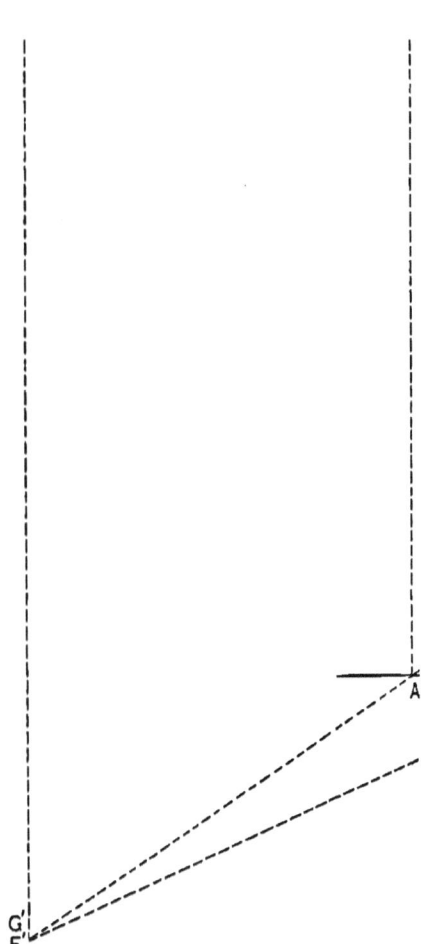

PLATE VII.

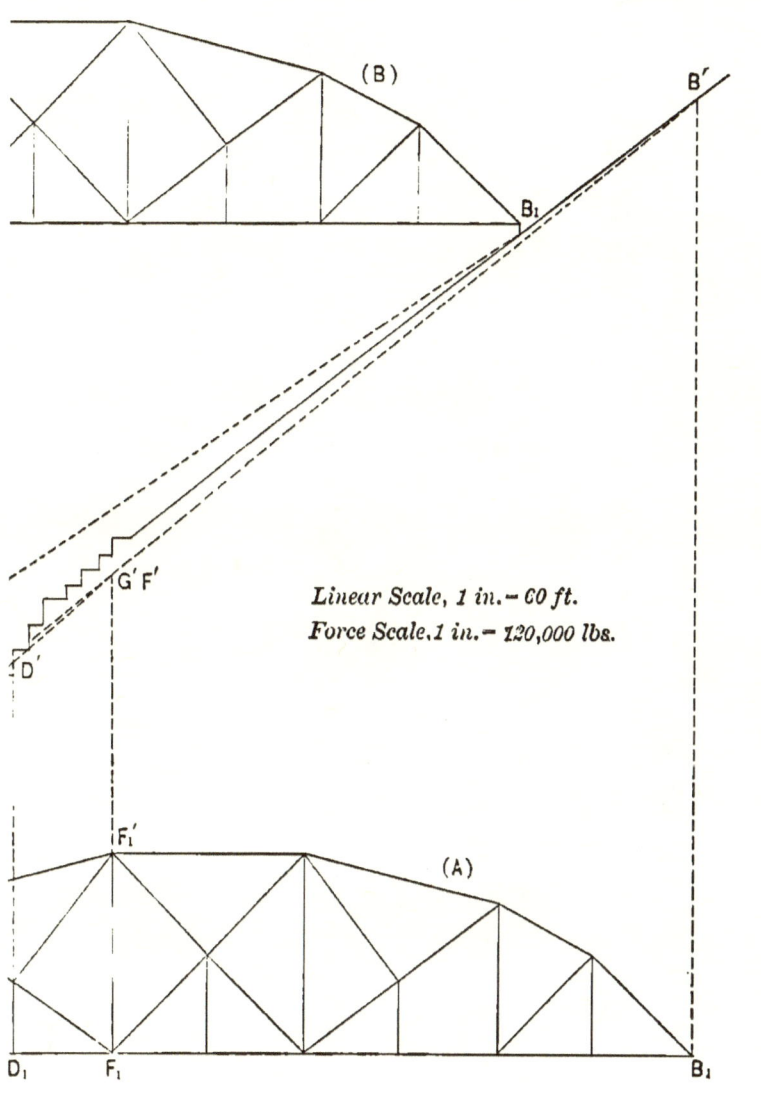

Linear Scale, 1 in. = 60 ft.
Force Scale, 1 in. = 120,000 lbs.

Fig. 48

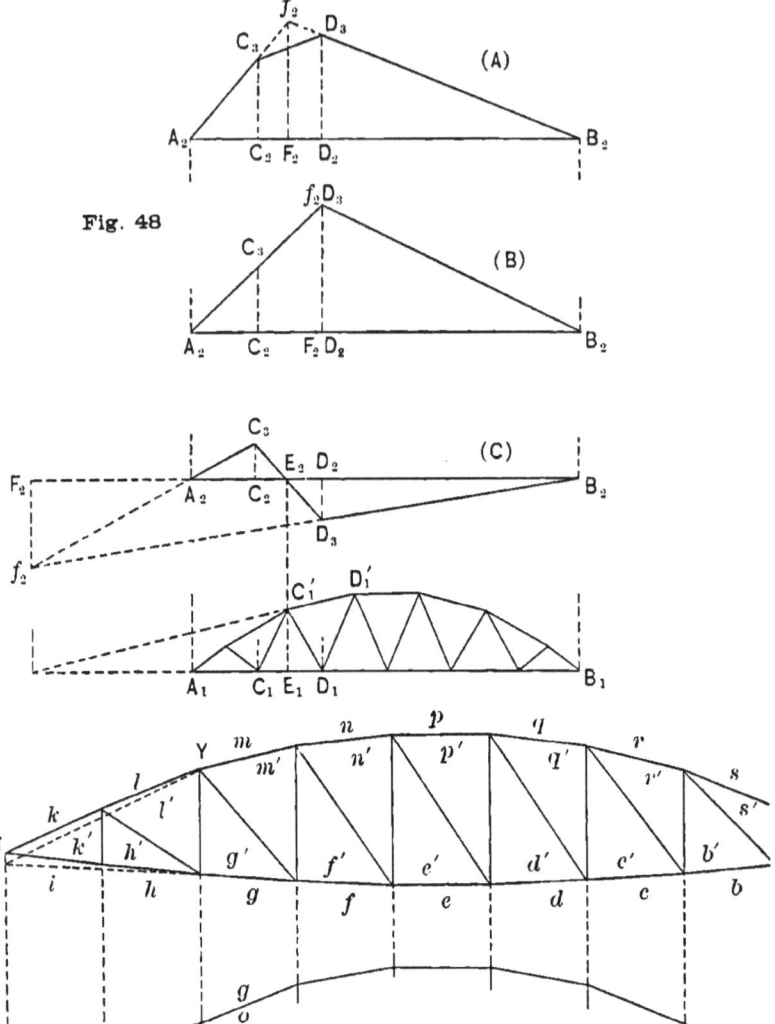

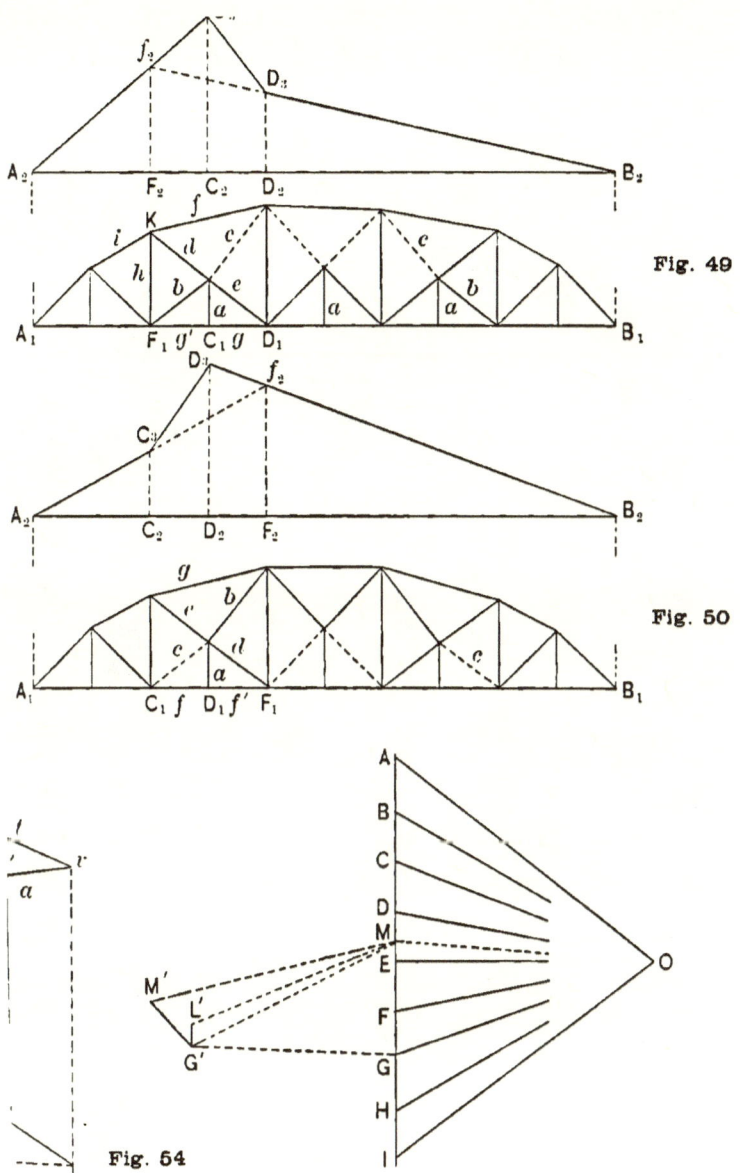

Fig. 49

Fig. 50

Fig. 54

www.ingramcontent.com/pod-product-compliance
Lightning Source LLC
Chambersburg PA
CBHW021802230426
43669CB00008B/604